D1618557

MODERN ASPECTS OF PROTEIN ADSORPTION ON BIOMATERIALS

58. 1 / 1. 59

Modern Aspects of
Protein Adsorption on Biomaterials

Edited by

Y.F. MISSIRLIS

and

W. LEMM

Berliner Hochschule für Technik
Campusbibliothek
– ausgesondert –

15,34 £

Technische Fachhochschule Berlin
Bibliothek C 30,–
Forum Seestrasse
Seestraße 64
1000 Berlin 65

BIO 03/69

KLUWER ACADEMIC PUBLISHERS
DORDRECHT / BOSTON / LONDON

Library of Congress Cataloging-in-Publication Data

Modern aspects of protein adsorption on biomaterials / edited by Y.F.
Missirlis and W. Lemm.
    p.   cm.
  Based on papers presented at a workshop held in Patras, Greece, in
June 1989.
  Includes bibliographical references and index.
  ISBN 0-7923-0973-1 (HB : alk. paper)
  1. Blood proteins--Absorption and adsorption--Congresses.
2. Biomedical materials--Congresses. 3. Biocompatibility-
-Congresses. I. Missirlis, Y. F., 1946-
QP99.3.P7M63 1990
612.1'18--dc20   90-5339

ISBN 0-7923-0973-1

Published by Kluwer Academic Publishers,
P.O. Box 17, 3300 AA Dordrecht, The Netherlands.

Kluwer Academic Publishers incorporates
the publishing programmes of
D. Reidel, Martinus Nijhoff, Dr W. Junk and MTP Press.

Sold and distributed in the U.S.A. and Canada
by Kluwer Academic Publishers,
101 Philip Drive, Norwell, MA 02061, U.S.A.

In all other countries, sold and distributed
by Kluwer Academic Publishers Group,
P.O. Box 322, 3300 AH Dordrecht, The Netherlands.

C 372192

Printed on acid-free paper

All Rights Reserved
© 1991 by Kluwer Academic Publishers
No part of the material protected by this copyright notice may be reproduced or
utilized in any form or by any means, electronic or mechanical,
including photocopying, recording or by any information storage and
retrieval system, without written permission from the copyright owners.

Printed in The Netherlands

Table of contents

Foreword   vii

List of contributors   ix

PART ONE: METHODS AND TECHNIQUES TO STUDY THE BLOOD
PROTEINS – ARTIFICIAL SURFACES INTERACTIONS

1. Can protein adsorption studies using radioisotopic techniques be
   useful for the development of hemocompatible materials?
   Ch. Baquey, F. Lespinasse, J. Caix, A. Baquey, B. Basse-Chathalinat   3

2. Competitive adsorption of proteins on polymer films: In situ
   measurements
   A. Baszkin   19

3. Protein adsorption and the interaction of human endothelial cells
   with polymer surfaces
   A. Dekker and T. Beugeling   25

4. Protein adsorption in relation to platelet adhesion and contact
   activation
   A.A. Poot and T. Beugeling   29

5. Studies of protein adsorption relevant to blood compatible
   materials
   J.L. Brash   39

6. Methods to study blood-surface interactions
   J.P. Cazenave and J.N. Mulvihill   49

7. Protein adsorption at polymer-liquid interfaces using series of
   polymers with varying hydrophilicity, charge and chain mobility
   H.S. van Damme and J. Feijen   55

8. Cooperative protein adsorption on surfaces with controlled
   alkyl-residue lattices
   H.P. Jennissen   63

9. Protein adsorption tests for polymer surfaces
   W. Lemm   73

10. Advantages and problems using FT–IR spectroscopy to study blood- surface interactions by monitoring the protein adsorption process
A. Magnani, M.C. Roncolini and R. Barbucci   81

11. Biocompatibility research at Humboldt University
H. Wolf   87

PART TWO: DISCUSSION CHAPTERS

12. Protein characterization   95

13. Labelling and protein identification techniques   113

14. Problems of protein adsorption studies – Lyman hypothesis   137

15. Polymer surface properties   149

16. In vitro, in vivo or ex vivo studies   163

17. Kinetics of protein adsorption   169

18. The Vroman effect   201

19. Protein adsorption and thrombus formation   219

20. Concluding remarks   249

Authors' index   257

Subject index   259

Foreword

The present book relates to the scientific records of a workshop held in Patras, Greece, in June 1989, under the auspices and with financial support of the European Economic Communities (Concerted Action EUROBIOMAT – Hemocompatibility – of the Medical Research Programme, Project: II.1.2/2). This concerted action promotes the collaboration on science and technology on the particular field of hemocompatible biomaterials: exchange of experts, scholarships and scientific workshops within the EC-member countries and COST countries such as Sweden, Finland, Turkey, Switzerland.

The first part of this monography refers to the oral presentations of the participants. The second part gives the book its unique character: the scientific discussion on updated aspects of protein adsorption of synthetic polymers in contact with blood. This second part is subdivided into nine chapters where specific topics were discussed freely, open-minded and even controversially.

This book intends to elucidate recurrent questions concerning the initial event when blood contacts artificial surfaces. Young investigators will consider this book to be appropriate to get familiar with the scientific background and the most relevant techniques and methods.

The international expert group of polymer chemists, physical chemists, hematologists, biologists, chemical engineers, provided an excellent exchange of ideas and arguments with respect to this highly interdisciplinary area, the interaction of plasma proteins with biomaterial surfaces.

List of contributors

Dr. Ch. Baquey
INSERM U. 306/CEEMASI
University of Bordeaux
146 Rue Léo Saignat
33076 Bordeaux Cedex
France

Dr. A. Baszkin
URA CNRS 1218
Physico-Chimie des Surfaces
University of Paris
5 Rue Jean-Baptiste Clément
92296 Chatenay-Malabry Cedex
France

Dr. T. Beugeling
Department of Chemical Engineering
University of Twente
P.O. Box 217
7500 AE Enschede
The Netherlands

Prof. J. Brash
Department of Chemical Engineering
McMaster University
L8S 4L7 Hamilton, Ontario
Canada

Prof. J.P. Casenave
CRTS de Strasbourg
10 Rue Spielmann
67085 Strasbourg Cedex
France

Dr. H.S. van Damme
Department of Chemical Engineering
University of Twente
P.O. Box 217
7500 AE Enschede
The Netherlands

Prof. A. Hoffman
Bioengineering, FL-20
University of Washington
98195 Seattle, Washington
USA

Prof. H. Jennissen
Institute for Physiological Chemistry
University Clinic Essen
Hufelandstrasse 55
4300 Essen 1
Germany

Prof. P. Koutsoukos
Department of Chemical Engineering
University of Patras
26100 Patras
Greece

Dr. W. Lemm
Rudolf Virchow Clinic
Spandauer Damm 130
Charlottenbourg
D-1000 Berlin 19
Germany

Dr. A. Magnani
Department of Chemistry
University of Siena
Pian dei Mantellini 44
53100 Sienna
Italy

Prof. Y. Missirlis
Laboratory of Biomedical Engineering
University of Patras
26100 Patras
Greece

Prof. H. Wolf
Humboldt University
Department for Stomatology and Biomaterial
Tucholskystrasse 2
1040 Berlin
Germany

PART ONE

Methods and techniques to study the
blood proteins – artificial surfaces interactions

1. Can protein adsorption studies using radioisotopic techniques be useful for the development of hemocompatible materials?

CH. BAQUEY, F. LESPINASSE, J. CAIX, A. BAQUEY* and
B. BASSE-CHATHALINAT

INSERM U. 306 et Laboratoire de Biophysique, Université de Bordeaux II 146, Rue Leo Saignat, 33076 Bordeaux Cedex, France
* Laboratoire d'Immunologie Cellulaire, Université de Bordeaux II, 146 Rue Leo Saignat, 33076 Bordeaux Cedex, France

For every people who try to prepare biocompatible materials and more particularly hemocompatible materials, the availability of experimental methods for the evaluation of their hemocompatibility is still of concern. This may appear as a paradox if one considers the great number of papers dealing with this subject but very few among them, maybe none of them, do answer to the question of the evaluation of the functional hemocompatibility of a material. In other words, it is very difficult to find one experimental method beside chronic in vivo experiments, so far as models used are valid, which may give results allowing any prediction about the behaviour of a material for a given clinical application.

The aim of this paper is to present our contribution to this search for methodologies and more especially to discuss advantages and limits of radiotracers based methods which are currently used in our laboratory.

Radiotracers, having the same physical chemical and biological properties as their non labelled counterparts, allow the experimenter to be aware of the fate of the latter, inside more or less complex experimental systems or living organisms thanks to non-invasive (i.e. mechanically independent of the system under investigation) means for detection and measurement. These methods are well adapted to study the adsorption of proteins onto materials, or more exactly protein retention phenomenon, adsorption being only one of the events which may occur, responsible for such phenomenon. This paper will deal with such studies, which are carried out either in vitro or in vivo.

In vitro studies may fulfill one or several of the following objectives:
- to measure the affinity for a given protein of various samples belonging to the same material family but corresponding to different compositions;
- to check that samples corresponding to a unique material composition, but belonging to different production batches are identical;
- to control the quality of a production process through the evaluation of the homogeneity with which a superficial characteristic is expressed;
- screening of materials in order to decide of the opportunity of their involvement in ex-vivo experiments.

Ex-vivo studies are aimed at the direct study of the interaction of flowing

3

Y.F. Missirlis and W. Lemm (eds), Modern Aspects of Protein Adsorption on Biomaterials, 3-18.
© 1991 *Kluwer Academic Publishers. Printed in the Netherlands.*

4

blood, free of any anticoagulant, with previously selected materials, this selection being made through in vitro studies which have been already quoted.

Material and methods

Reference material

PVC tubing: Surgical grade tubing (I.D = 4 mm) is provided by 3M (Saint-Paul, Minnesota) under the reference S50-KL (lot #9) for the IUPAC International Cooperative Project dealing with testing of NHBLI* primary reference material.

Polyethylene tubing: Medical grade tubing (2.9 × 4 mm) is obtained from Clay Adams (a Division of Becton & Dickinson, Parsippany NJ, USA) and rinsed extensively for 2 hours with circulating Michaelis' buffer (10 ml.mn^{-1}).

Silicon elastomer tubing: Medical grade tubing (3×5 mm and 2×4 mm) is provided by SODIP (a Division of HOSPAL, Meyzieu, France) under the trademark SCURASIL.

Tentative material

'Heparin-like' materials: Polyethylene tubing corresponding to the quality which has been just described here above is especially treated according to an original procedure (Migonney et al. 1986 (Ref. 14)) at the Laboratoire de Chimie des Macromolecules de l'Université Paris Nord (France) in order to give it a heparin-like ability to catalyse the inhibition of thrombin by antithrombin III.

Polyurethane tubing (PUT): Tubing (3 × 5 mm) made of aliphatic poly-etherurethane elastomers, the main originality of which comes from the compound bringing the isocyanate groups, is provided by the Institut du Pin from the University of Bordeaux I (France).

Carbon-Carbon Composites: The composites used have been prepared by the Société Européenne de Propulsion (Etablissement de Bordeaux, France). Two types of materials were available; both had a matrix made of pyrolytic carbon for 85% and of silicon carbide for 15%, but the organization of the carbon fibers was different and gave a smaller porosity for material A than for material B. Moreover materials could get either of two different surface treatments F1 and F2 after machining. For the present study (must be suppressed in front of 2.5 cm long tubular samples (3.5 × 4.5 mm) were machined; their wall at each

---

* NHBLI: National Heart Blood and Lung Institute (Belhesda, USA)

end was carefully sharpened so as to diminish turbulence occurrence when fluids would be allowed to flow through them.

Gelatin-coated vascular grafts: Knitted Dacron® vascular prostheses which are coated with crosslinked gelatin are supplied by the Laboratoires Bruneau, Boulogne, Billancourt (France).

Table 1. A very coarse chemical structure of heparin-like and control surfaces under test for the adsorption of antithrombin III and typical data obtained during such a test.

	Tubes tested						
	A			B			C
	$PE-PS\begin{cases}SO_3- \\ SO_2\text{-}ASP\end{cases}$			$PE-PS-SO_3-$			PE
Retained concentration of AT III* in pmol/cm² of inner surface	67.3	69.6	69.6	60.3	67.3	64.9	18.6 23.2 20.8
Retained concentration of anti AT III* antibody in pmol/cm² of inner surface	69.6	69.6	71.9	6.9	9.2	6.9	0

PE = Polyethylene
PS = Polystyrene

Table 2. Kinetics parameter of platelet adhesion on heparin-like modified polyethylene and control polyethylene surfaces during ex-vivo circulation. (s_p ± s.e.m.) cells.mm⁻².s⁻¹.

Contact duration with flowing blood	t ≤ 20 mn	20 ≤ t ≤ 45 mn	45 ≤ t ≤ 130 mn
Polyethylene	163 ± 101 $n = 4$		
Modified polyethylene	17.6 ± 11 $n_1 = 7$	51.6 ± 28 $n_2 = 7$	87.3 ± 31 $n_3 = 7$

Biological fluid

For in vitro experiments buffered solutions of a single protein and decalcified plasma are used. Decalcified plasma which is pooled from several donors and kept at 4°C, is collected by the regional blood bank (Centre de Transfusion Sanguine, CTS, Bordeaux, France). Protein suppliers are mentioned later on, in the paragraph dealing with radiotracers.

6

Animals

Mongrel dogs weighing more than 25 kg are fasted for 24 hours before surgery. An hematologic check up is carried out during this period providing blood cell numerations, plasma fibrinogen concentration, prothrombin level and clotting times (activated partial thromboplastin time and thrombin time).

Flow generators

For in vitro studies such set ups are designed to allow one of the previously described biological fluids to circulate, in contact with the biomaterial surfaces of interest. The working part may be a peristaltic pump which gives rise to a slightly pulsatile flow of which physical characteristics are by no means equivalent to those of the physiological blood flow.

In order to get flowing conditions much more similar to physiological ones we may use a more sophisticated set-up (Fig. 1). Tubular samples of the material of interest are included in a so-called secondary circuit through which the biological fluid goes from a ventricle equivalent through an aortic valve equivalent and back to this ventricle through a mitral valve equivalent. The ventricle is hydraulically activated according to a pattern which is chosen to obtain previously established characteristics for the pressure and flow rate waves inside the sample of interest. This pattern is defined by the working mode of the pump responsible for the circulation of the fluid through the primary

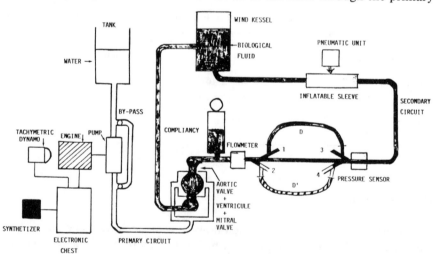

Fig. 1. Hydrodynamic generator adapted from an original device which has been developped at the Institut de Mechanique de Fluides in Marseille (France);
D: standarization loop
D': measurement loop
1-2-3-4: special connectors avoiding turbulences to occur.

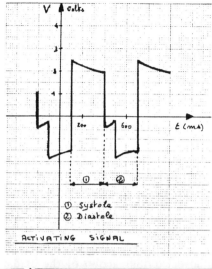

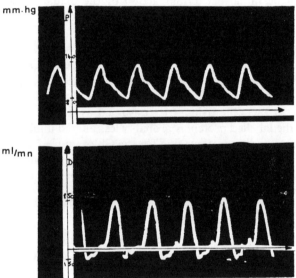

Fig. 2. These graphs show a possible shape for the activating signal of the pumping unit (see Fig. 1) and the corresponding pressure and flow rate waves recorded for the fluid in the secondary circuit (see Fig. 1).

circuit, this working mode resulting from the shape of a triggering signal given by a synthesizer. Thus playing with the setting knobs of the 48 potentiometers which fit this synthetizer, allows a correct fitting of the ventricule activation pattern (Fig. 2).

RADIOTRACERS

Albumin

Human albumin conditioned under the reference TCK2 by ORIS Industrie S.A. (B.P. 21, Gif sur Yvette, France) for labelling with [99mTc] from sodium pertechnetate, was used as well as radio-iodinated albumin prepared in our laboratory according to the following techniques.

Chloramine T method

The oxidation method of iodine by chloramine T (N-chloro-p.toluene sulfonamide) had been first proposed by Greenwood et al. (Ref. 1). The chloramine T gets its oxidation power from hypochloric acid, which is released in solution, and, hence, changes the iodide ions into active forms.

It is a quick method giving a good labelling yield (30 to 40%) but as the chemical reactions are developed in homogeneous phase, the native substratum is exposed to the initial deleterious action of the oxidant, and the labelled product to the final de-iodinating action of the reducing agent.

Lactoperoxydase method

The lactoperoxydase catalyses the oxidation of iodide ions in the presence of hydrogen peroxide simultaneously produced in situ (Ref. 2, 3).

It is usually admitted that this technique is better for the labelling of fragile polypeptides. In fact, the oxidant is consumed as it appears in the reactive medium and the labelled protein is probably less altered under such conditions than it is under conditions prevailing for the Chloramine T method.

However, the labelling yield is often lower, between 10 and 30% on account for the competition for iodine, existing between the protein of interest and the other proteins present in the mixture.

Iodo-Gen method

Iodo-Gen (Iodo-Gen™ 1.3.4.6. tetrachloro 3 6 diphenyl glycoluril. Pierce and Warriner U.K. Ldt., Chester, England or Co Box 117 Rockford, Illinois 61105) is an organic molecule known since 1910, used by Fraker and Speck, (Ref. 4) and more recently by Salacinski et al. (Ref. 5).

Operating in heterogeneous phase which prevents the labelled protein being exposed to the generating agent of active Iodine, seems to be particularly interesting. Moreover, labelling yields can reach 50%.

Fibrinogen

Human fibrinogen labelled with [125I] was obtained from Amersham-France (Avenue du Canada, B.P. 144, 91944 Les Ulis-Cedex France) under the reference IMS 30. Each vial contains 110 μC (4.07.10^6 becquerels). For in vitro studies 0.8 mg are reconstituted with 11 ml of sterile water. This solution is isotonic as it contains sodium chloride (0.65%), sodium citrate (0.75%) and human albumin (22 mg). It must be stored below 4°C or used immediately.

For in vivo studies, either canine fibrinogen, purified from a pool of canine plasma collected from several animals, or human fibrinogen (KABI) are labelled according to methods very similar to those used with albumin.

The only difference with the classical Chloramine T method comes from the suppression of the reduction step which would destroy the fibrinogen disulfide bonds. To isolate the labelled protein from excess of oxidative agent and Na*I, the reacting mixture is directly put on a gel permeation chromatography using Sephadex G50 as recommended by Okada and Blomback (Ref. 6).

Antithrombin III

AT III of human origin is isolated and purified by the blood bank in Lille, CTS-Lille, France (special batch No. 78302). Each vial contains 600 U to be made up in 20 ml distilled water. Iodination is carried out according to a Iodo-Gen technique somewhat related to the method described here above for albumin, but at 4°C because the protein is very fragile and sensitive to variations in temperature (Ref. 7).

The mixture resulting from the labelling steps is loaded onto a column filled with Heparin Ultrogel A4R (– IBF – Biotechnics, 92390 Villeneuve la Garenne, France) for affinity chromatography.

A first elution is performed with a phosphate buffer at low ionic strength (NaCl 0.4 M) which releases a protein fraction without any affinity for heparin on the one hand, and iodide ions in excess on the other. Then a second elution performed with a phosphate buffer at high ionic strength (NaCl 1.5 M) releases the protein fraction with high affinity for heparin corresponding to a high specific activity.

MEANS FOR RADIOACTIVITY MEASUREMENT AND IMAGING

Detectors based on junctions of semi-conductors (HPGC i.e. Highly Pure Germanium or GeLi) are preferred because of their good energy resolution (− 1.5%). Such a performance is necessary when several isotopes with proximal photon energy emissions are used as we do for ex-vivo experiments (Ref. 8). The detector is connected to a spectrometric analysing system and a recording device, data being processed by an on-line microcomputer.

For imaging purposes, a scintillation camera (Acti-Camera CGR) is used to determine the amount of an adsorbed protein on a given surface and to study its distribution on this surface from the radioactivity recorded for the corresponding tracer. The detection is made by a large crystal of NaI (dopped with Thallium), the size of which defines the limits of the explored field (a few 10 cm in diameter).

The main physical characteristics of such a detector are: an energy resolution of about 15% and a spatial resolution of about 5 mm (Ref. 9, 10). The first characteristic limits the use of the detector to only two simultaneous tracers, as far as their respective energies differ by more than 15%. The second characteristic allows images to be obtained with a resolution no better than 5 mm, this giving the macroscopic distribution of the tracers. The images can be quantitatively processed by an associated computing system.

If a spatial resolution far better than that of a scintillation camera is required, an autoradiographic method can be used.

The principle of these well known techniques consists in placing the radiolabelled surface in contact with an electron sensitive photographic emulsion which acts as a detector. After a suitable period of contact the image is developed and the autoradiographic document contains the tracer distribution on the surface of interest. This distribution is examined with a light microscope with a resolution of about one micrometer (Ref. 11).

EXPERIMENTAL PROCEDURE USED TO STUDY THE INTERACTION OF PLASMA PROTEINS WITH BIOMATERIALS

In vitro

Material samples are thoroughly rinsed with Michaelis' buffer, and then immersed in the biological fluid for 30 minutes. At the end of this incubation period samples are rinsed with saline. Each sample is put into a vessel feeded with saline and designed to keep a constant level of fluid. The rinsing stage is stopped when the radioactivity of the liquid out of the rinsing vessel becomes of the same order of magnitude as the background.

Most of the time, samples (test and control) available as tubing segments are mounted in series in a closed circuit through which a biological fluid is allowed to flow for half an hour after several pre-incubation periods which are carried out according to a procedure already described elsewhere (Ref. 12) with circulating fresh Michaelis' buffer. At the end of the last pre-incubation period, the buffer is replaced by the biological fluid of interest, i.e. decalcified plasma diluted to the tenth for the fibrinogen or albumin adsorption studies, or buffered solutions of albumin fibrinogen or antithrombin III when the affinity for this protein of 'heparin-like' materials has to be studied. Then an aliquot of a buffered solution of labelled albumin, or fibrinogen or antithrombin III as well, is introduced into the circuit. Generally, operations are lead at room temperature but they can be lead also at 37°C.

Whereas the biological fluid circulates, the whole radioactivity of a given length (involving the material of interest) of the circuit is measured. The recorded value Rp (Fig. 3) corresponds to the whole amount of protein Q_T

(should it be bound to the tubing wall or free in the bulk of the circulating medium) which was initially available for adsorption. This dynamic stage of incubating is followed by a rinsing stage with saline. The recorded radioactivity decreases steeply and then more and more slowly, until a minimum value Rw, which corresponds to the amount of protein bound to the tubing wall, is reached. We have already shown (Ref. 12, 13) that the superficial concentration Cs of the bound protein is given by the following expression:

$$Cs = (C*r/2)*(Rw/Rp)$$

where C is the initial bulky concentration of the protein of interest, and r the inner radius of the tubing.

The rinsing stage can be either followed by a more severe rinsing stage with a 3M NaCl solution, or by an incubation stage with the same biological fluid free of radiolabelled protein in order to look at the ability of the bound protein to exchange with its bulky equivalent. When this exchange occurs, the bound protein is said to be pseudo-reversibly adsorbed. It may be interesting to compare the percentage of bound protein which can exchange to that which can be released at high ionic strength.

In case of the study of the affinity of antithrombin III for 'heparin-like' materials, and in order to get some information about the conformational state of this protein when it is adsorbed, its affinity for a monoclonal antibody which uses to bind specifically to this protein in solution, can be checked. In this respect, the rinsing stage is followed by a further dynamic incubation stage with a buffered solution of an anti AT III monoclonal antibody labelled with Indium 111 by a coupling agent (DTPA). The amount of bound antibody to the adsorbed antithrombin III present on the surface of interest, is assessed using

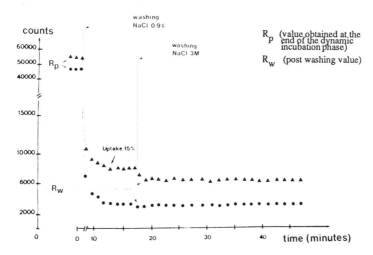

Fig. 3. In vitro study of the adsorption of Antithrombin III onto heparin-like tubing (▲) on the one hand and onto medical grade polyethylene (•) on the other.

a procedure similar to that which has been described to determine the amount of a given protein directly bound to a surface.

Ex vivo

The blood of an anaesthetized dog is derived through an extracorporeal derivation made of the material to be tested and going from an artery to a vein (A-V shunt) or to downstream the same artery (A-A by-pass). Before blood is allowed to flow through the derivation, the animal received several radiotracers, radio-iodinated fibrinogen or radio-iodinated albumin which have been previously described, but also autologous ^{111}In labelled platelets and autologous ^{99m}Tc labelled red blood cells, (^{99m}Tc-RBC) because this experiment is also aimed at the study of the interaction of blood cells with materials and more generally of thrombogenic phenomena.

Quantitative analysis of the radioactivity of periodically collected blood samples, gives information about the respective evolution of the concentration of these tracers in the whole circulation after their intravenous injection. When the curves related to these evolutions show a constant slope, blood is allowed to flow through the derivation. This one is settled on the bottom of a lead walled container surrounding the detector.

Each of these tracers contributes to the whole volumic radioactivity of whole circulating blood, in proportion to the concentration of the related biological species. As a matter of fact each of these tracers contributes also to the whole radioactivity of a given length of the extracorporeal derivation in proportion to the amount of the related biological species contained, at the time of measurement, either in the flowing blood or as deposited material onto the wall of the conduit. A comparison of these contributions to the circulating blood radioactivity on the one hand, and to the A-V shunt or A-A by-pass radioactivity on the other, may show a difference which obviously comes from the presence of radioactive material on the wall of the derivation.

Which type of information is assessable?

Analyzing several examples will help to answer this question: *Comparison of several polyurethane based elastomers for their respective affinities for albumin and fibrinogen.*

Results obtained under static conditions dealing respectively with albumin adsorption and fibrinogen adsorption are reported on Fig. 4 and Fig. 5. More interesting are the figures obtained when the ratio of the amount of adsorbed fibrinogen to the amount of adsorbed albumin is calculated. These values are reported on Fig. 6 allow to distinguish two classes of materials:
– a class of materials of which this ratio is greater than or equal to 0.85 (0.85 $\leq p \leq 1.18$)

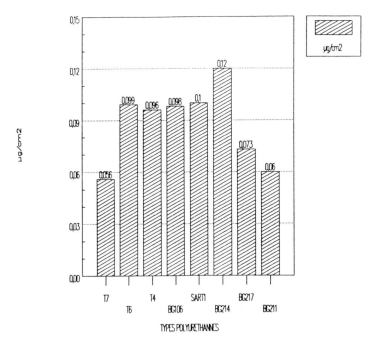

Fig. 4. Static in vitro study of the adsorption of fibrinogen onto several flat samples of polyurethane-based materials, from plasma diluted to the tenth.

– a class of materials for which this ratio is twice smaller ($0.32 \leq p \leq 0.60$).

In general materials which have a greater affinity for fibrinogen than for albumin are said to be more thrombogenic. Our ex-vivo experiments should confirm this saying, but at this moment preliminary results show that materials related to the reference BG 106 demonstrate in vitro an affinity for platelets which is equivalent or smaller than that of SCURASIL, and an ex-vivo patency which is equivalent or better than that of SCURASIL.

QUANTITATIVE AND QUALITATIVE COMPARISON OF THE ANTITHROMBIN III
AFFINITIES OF POLYETHYLENE BASED MATERIALS MODIFIED ACCORDING TO
THE 'HERAPIN-LIKE' PROCEDURE

According to their production batch, these materials may be characterized by different substitution degrees, which means by different superficial densities of chemical groups such as -COOH, $-SO_2-NH-R$ or $-SO_3H$. Materials coming from the same batch have been shown (Ref. 13) to have similar affinities for antithrombin III, but in the same paper we reported also an equivalent affinity for a completely sulfonated polyethylene-polystyrene grafted copolymer which has been shown to be rather thrombogenic in vivo.

So it can be concluded that the superficial chemical structure of such material

ADSORPTION PROT.

ALB.131.I.

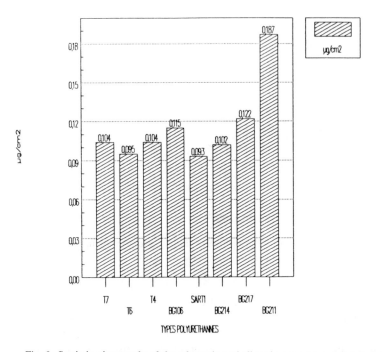

Fig. 5. Static in vitro study of the adsorption of albumin onto several flat samples of polyurethane-based materials from plasma diluted to the tenth.

is without influence on the whole amount of antithrombin III able to bind to their surface, but can play a role on the ability of bound protein molecules to express their specific biological functions.

In order to check this ability, the affinity for these bound molecules of the monoclonal antibody directed towards ATIII, has been measured and found very small when antithrombin is adsorbed on the sulfonated polyethylene polystyrene copolymers, while a mole-to-mole binding is observed between the antibody and antithrombin III when the latter is adsorbed on the so-called 'heparin-like' materials (Ref. 13).

We may suppose that ATIII suffers conformational changes, when it adsorbs to the sulfonated polyethylene-polystyrene grafted copolymer, which could explain that the antibody is unable to recognize its related antigen. As a matter of fact, 'heparin-like' materials coming from different production batches may demonstrate the same affinity for antithrombin III without having the same

Adsorption PROTEINES

RAPPORT fib/alb

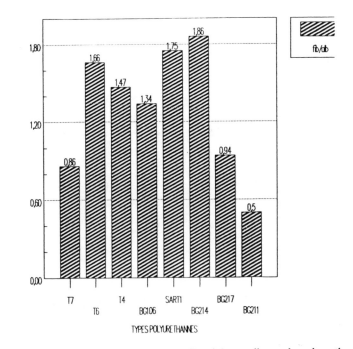

Fig. 6. Evolution of the ratio [Fibrinogen]/[Albumin] according to the polyurethane-based material which is considered (values are not coming from the same experiment as in Fig. 4 and 5).

superficial chemical structure which seems to play the main role for their antithrombic activity. We observed indeed in ex-vivo experiments some discrepancies between data obtained with materials coming from different batches. Accordingly, antithrombin III adsorption studies must be completed by other tests in order to control the quality of such materials.

QUANTITATIVE EX-VIVO STUDY OF FIBRINOGEN DEPOSITION ON DIFFERENT MATERIALS

For several materials the contribution of the labelled fibrinogen to the whole radioactivity of the extracorporeal circuit (ECC) and the contribution of this tracer to the radioactivity in the whole circulation are decreasing with the same negative slope; that means there is no detectable fibrinogen retained by the tubing wall.

16

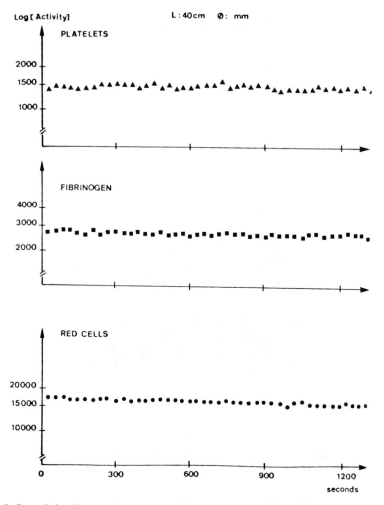

Fig. 7. Recorded radioactivities related to each of the involved tracers during extracorporeal circulation of blood through a polyethylene tubing modified according to the 'heparin-like' procedure.

For other materials the two slopes which have just been considered are different as for instance on Fig. 7, (but we find analogous data for carbon-carbon materials) where the slope of the fibrinogen contribution to the whole radioactivity of the ECC is almost zero. From the recorded curve the amount of fibrinogen or fibrin deposited onto the wall has been calculated and found equal to 15 $\mu g/cm^2$ after a 15 minutes exposure, which is much more important than a monolayer. During the same period of time the contribution of the labelled platelets to the whole radioactivity of the ECC remains stable as it does in the whole circulation. Then two hypotheses can be proposed:
i) platelets are not adhering to the wall of the ECC neither to the fibrinogen or fibrin deposit,

ii) platelets are adhering to the wall of the ECC but very soon after blood is allowed to flow through the circuit. This adhesion phenomenon could last no more than 30 to 60 seconds; that's the reason why it is hardly detectable taking into account the time constant of the measurement sequence which is of the same order of magnitude. The resulting platelet carpet would provide a high affinity substrate for fibrinogen adsorption, but it is difficult to explain why platelets would stop to adhere.

Conclusion

As far as in vitro protein adsorption studies are concerned, the only meaningful advantages of our methods, compared to those used by others, come from the available means for radioactivity measurement and imaging. These ones allow us to involve two radiotracers in the same experiment, and even more, if overlapping of emission spectra is not too large. So competitive adsorption studies can be easily carried out.

Scintigraphic imaging at a macroscopic level as well as autoradiographic imaging at a microscopic level, appears as good tools to reveal homogeneity defects for the material surfaces whether these defects are morphological or coming from an heterogeneity of the local chemical structure. Lastly the use of radiolabelled monoclonal antibodies directed towards the proteins we are interested in, offers a valuable opportunity to characterize in situ the conformational state of an adsorbed protein, and its eventual functionality.

The unique advantages or our ex-vivo method which is currently used in our laboratory lies on its ability to make possible a non-invasive follow-up of phenomena occurring at a flowing blood-material interface. As the radioactivity corresponding to several biological species is continuously analyzed and recorded, kinetics of their eventual deposition can be easily assessed and for instance eventual embolization of platelet aggregates can be detected. All other methods proposed to study the interaction of flowing blood with biomaterials permit at the best to establish the status of the exposed surface from time to time, but don't bring any information about events responsible for status changes between two observation times.

In our opinion several imperatives must be taken into account to expect methods described here be the most performant.

Firstly quality control of radiotracers must be given the priority. Labelled proteins must keep their original electrophoresis pattern as shown for fibrinogen (Ref. 12), or their original affinity for a specific ligand as shown for antithrombin III. Labelled cells must keep their physiological properties: same biological half-life, same functionality (platelets for instance keep their ability to aggregate in the presence of ADP or collagen...)

Secondly, figures obtained for amounts of adsorbed proteins or for densities of adhered platelets on a given material must not be considered as absolute values but weighed by corresponding values obtained with one or several

18

reference materials during the same experiment. That is particularly important for ex-vivo experiments.

References

1. 'The preparation of ^{131}I labelled human growth hormone of high specific radioactivity', Greenwood F.C., Hunter W.M. & Glover J.S. (1963), *Biochem. J.* 89, 114-23.
2. 'An enzyme method for the trace iodination of immunoglobulins and other proteins', Marchalonis J.J. (1969), *Biochem. J.* 113, 299.
3. 'Lactoperoxidase coupled to polyacrylamide for radioiodination of proteins to high specific activity', Thorell J.I. & Johannson B.G. (1974), *Immunochemistry* 11, 203-6.
4. 'Protein and cell membrane iodinations with a sparingly soluble chloroamide, 1,3,4,6-tetrachloro-3a, 6a-diphenylglycoluril', Fracker P.J. & Speck J.C. (1978), *Biochem. Biophys. Res. Commun.* 20, 849-857.
5. 'Iodination of proteins, glycoproteins and peptides using a solid-phase oxidizing agent, 1,3,4,6-tetrachloro-3a, 6a-diphenylglycoluril', Salacinski P.R.P., McLean C., Sykes J.E.C., Clement-Jones V.V. & Lowry P.J. (1981), *Anal. Biochem.* 117, 136.
6. 'Factors influencing fibrin gel structure studied by flow measurement', Okada M. & Blomback B. (1983), *Molecular Biology of fibrinogen and fibrin*, Ed. M.W. Mosesson and R.F. Doodlittle, published by the New York Academy of Sciences, Vol. 408, 233.
7. 'Conditions for radioiodination of antithrombin III rataining its biological properties', Caix J., Perrot Minnot A., Beziade A., Vuillemin L., Belloc F., Baquey Ch. & Ducassou D. (1987), *Int. Applied Rad. Isot.* Vol. 38, No 12, 1003-1006.
8. 'Direct in vivo study of flowing blood artificial surface interactions. An original application of dynamic isotopic techniques!', Basse-Cathalinat B., Baquey Ch., Llabador y., Fleury A. (1980), *Int. J. of Applied Rad. and Isot.* 31, 747-751.
9. 'Scintillation camera with multichannel collimators', Anger H.O. (1964), *J. Nucl. Med.* 5, 515.
10. 'Techniques experimentales: les examens scintigraphiques', Ducassou D., & Isabelle D. (1975), in: *Traite de Medecine Nucleaire-Bases Theoriques*, edited under leadership of G. Meyniel, Flammarion-Medecine-Sciences-Paris, 3-4, pp. 65-78.
11. 'High spatial resolution imaging of labelled blood cells or proteins on artificial materials', Baquey Ch., More N., Basse-Cathalinat B., Beziade A. & Ducassou D. (1986), 'Biological and Biomechanical Performance of Biomaterials', *Advances in Biomaterials*, Ed. P. Christel, A. Meunier and A.J.C. Lee, 6, 293-298.
12. 'Quantitative and qualitative radioisotopic studies of the affinity of biomaterial surfaces for blood components under dynamic circumstances', Baquey Ch., Bordenave L., Caix J., More N., Basse-Cathalinat B., Ducassou D. (1989), to be published in *Tests procedures for blood compatibility of biomaterials*. Edite dans le cadre du projet 'Biomateriaux EUROXY' de la CEE (à paraître).
13. 'Control and isotopic quantification of affinity of Antithrombin III for heparin like surfaces', Caix J., Migonney V., Baquey A., Fougnot C., Perrot Minnot A., Beziade A., Vuillemin L., Baquey Ch. & Ducassou D. (1988), *Biomaterials*, Vol. 9, January, 62-65.
14. 'Fonctionnalisation de la surface interne de materiaux tubulaires. Etude de l'inhibition de la thrombine par l'Antithrombine III à la surface de ces materiaux', Migonney V. (1986), Doctorat Thesis, Paris.

2. Competitive adsorption of proteins on polymer films: In situ measurements

A. BASZKIN
Physico-Chimie des Surfaces et Innovation en Pharmacotechnie URA CNRS 1218, Université de Paris-Sud, 5 Rue J.B.Clement, 92296 Chatenay-Malabry Cedex, France

Introduction

Although protein adsorption on polymer surfaces has been subject of a large number investigations during the past decade, the mechanism of the process remains still unclear and the organization of various adsorbed protein layers at the polymer-solution interface is not well known. There is still a lack of direct experimental evidence concerning reversibility of adsorption from single protein systems to polymers of different surface compositions and in particular of data concerning protein adsorption from binary or complex mixtures like plasma.

The ambiguity which prevails in the today's protein adsorption state of the art is primarily due to:

1. the differences in techniques and experimental conditions under which protein adsorption process is studied;
2. the differences in the degree of purity of used proteins;
3. missing information on the composition of the surface region (the outer few atomic layers) of a polymer sample while in contact with the adsorption medium.

It is the purpose of this communication to draw the attention of protein adsorption experts of this workshop to the advantages in using [14C] proteins for studying competitive protein adsorption.

Methodology

The experiments of adsorption/desorption of [14C] proteins at the polymer/solution interfaces are performed with the measuring device shown in Fig. 1A.

The gas flow counter measures the radioactivity and continuously displays it on a recorder as a function of time. To allow for the radioactivity (Ab) of the thin liquid layer close to the polymer/solution interface two techniques are used:

1. The curves of adsorption (counts/min) vs. time corresponding to different protein concentrations in solution, are extrapolated to the time zero and the radioactivities at t = 0 are taken as the Ab values.

Y.F. Missirlis and W. Lemm (eds), Modern Aspects of Protein Adsorption on Biomaterials, 19-23.
© 1991 *Kluwer Academic Publishers. Printed in the Netherlands.*

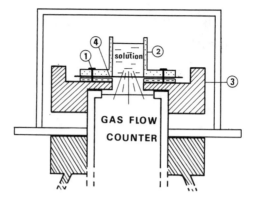

Fig. 1A. Adsorption/desorption measuring device. (1) Screws ensuring the tightness of the bottom of the cell. (2) Glass container. (3) Supports ensuring reproducibility of geometrical conditions. (4) Polymer film.

2. A separately run experiment is performed in which instead of a [^{14}C] protein the glass container is filled with a solution of a non-adsorbing substance containing the same radioactive element, e.g. [K^{14}CNS], and its radioactivity AB' is measured. The radioactivity Ab is then Ab = Ab'cp/c'p' were c and c' are concentrations of protein and non-adsorbing substance in solution, p and p' their respective specific activities.

Subtraction of Ab from the total measured radioactivity At gives the radioactivity corresponding to adsorbed protein molecules at the polymer-solution interface Aad (Fig. 1B) at each of the studied protein concentration in solution. The allowance for the absorption of radiation by each polymer sample has to be done before the adsorption experiment is performed. This is achieved with the help of [^{14}C] methyl methacrylate solid source placed above the polymer. Its radioactivity is measured in the same geometrical conditions as the adsorption measurements.

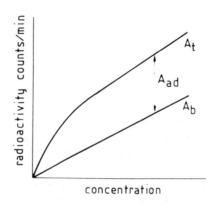

Fig. 1B. Schematic representation of the technique; A_t: total measured radioactivity; A_b: radioactivity corresponding to the volume phase; A_{ad}: radioactivity of the adsorbed protein.

The radioactivity Aad in counts/min is converted into mg/m² by depositing, drying on flat glass surfaces known amounts of [¹⁴C] protein which, when counted give a conversion factor of the amount of protein per unit area. Knowing the conversion factor and the area of the sample exposed to a protein solution the adsorbed protein amount in mg/m² is obtained.

To measure the in situ desorption at a given time, a protein solution is pumped out from the cell and simultaneously replaced by water, or a buffer solution. Multiple replacement cycles lead to a negligible protein concentration in the cell.

The loosely bound protein fraction (reversibly adsorbed protein) of the total adsorbed layer is thus obtained by subtraction of the irreversible adsorption from the total adsorption value (Fig. 2).

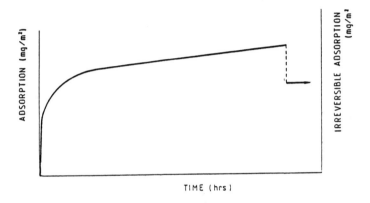

Fig. 2. A typical adsorption kinetics curve from a single protein sollution system. Dotted line and arrow indicate the amount adsorbed left after desorption.

Competitive adsorption

Competitive adsorption experiments in which adsorption of a given [¹⁴C] protein is studied from binary or complex mixtures of proteins are being realized with the same procedure as that used in adsorption experiments with single protein systems.

At the beginning of such an experiment the amount of molecules of each protein arriving to the polymer/solution interface is proportional to its concentration and diffusivity in solution. The displacement of the adsorbed protein by a protein of a higher surface affinity may take place in a latter stage of adsorption. At equilibrium of the exchange process the surface concentration of each protein would depend both on its solution concentration and its affinity to the surface. If in the first stage of adsorption the surface concentration of a protein attains a value higher than its value at equilibrium, then a lowering in adsorption would be observed in its kinetics curve.

The dependence of the appearance of [¹⁴C] collagen adsorption maximum on albumin concentration in solution is illustrated by the following example: At the constant solution concentration of collagen [coll] = 25 μg/ml, the appearance of the adsorption maximum is: at [alb] = 3.5 μg/ml – 25 min; at [alb] = 5 μg/ml – 15 min and at [alb] = 7.5 μg/ml – 4 min.

The problem of protein adsorption irreversibility has to be considered in light of competitive adsorption phenomena. Let us take, as an example, the data represented in Fig. 3.

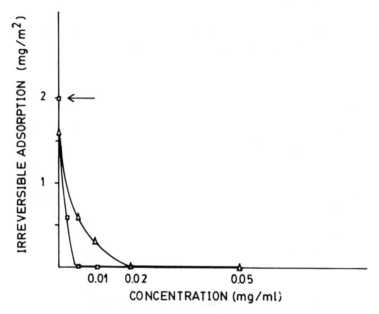

Fig. 3. Irreversible adsorbed collagen on polyethylene films in presence of albumin (□) or fibrinogen (Δ) in the adsorption solution. Collagen concentration: 2.5×10^{-2} mg/ml. Desorption after 20 hrs of adsorption with 0.2 M NaCl – 0.1 M CH$_3$COOH buffer adjusted with concentrated HCl to pH = 2.7.

The irreversibly adsorbed amount of collagen from the solution containing only collagen, on polyethylene films is indicated by an arrow. Although a relatively large amount of collagen is irreversibly adsorbed under these conditions, the presence of a second protein in solution considerably lowers this value. The presence of albumin in solution reduced this quantity to zero at the concentration higher than 5×10^{-3} mg/ml but 2×10^{2} mg/ml of fibrinogen in solution is needed to achieve the same effect.

The observed phenomena indicate that:
1. albumin affinity to the surface is higher than that of fibrinogen;
2. the irreversibly adsorbed quantities of collagen decrease when its adsorption is performed from a binary mixture of proteins.

The conformation changes associated with the irreversible adsorption in this case may in fact be hindered by the surface exchange process.

Concluding remarks

The advantages of using [^{14}C] labelled proteins for the in situ measurements of protein adsorption may be summarized as follows:
1. Kinetics of adsorption of a given protein from a single or multiprotein containing systems may be continuously measured;
2. Desorption of the loosely bound of adsorbed protein by a buffer solution may be performed at any time of the adsorption process. The reversibly/irreversibly adsorbed protein ratio may be known for each protein-polymer system;
3. Direct monitoring of the Vroman effect.

References

1. Baszkin, A., Deyme, M., Perez, E., Proust, J.E., 'Reversible-irreversible protein adsorption and polymer surface characterization', in: Brash J.L. & Horbett T.A. (Eds), ACS Symposium Series 343, American Chemical Society, Washington DC, 1987.
2. Deyme, M., Baszkin, A., Proust, J.E., Perez, E., Albrecht, G., Boissonnade, M.M., 'Collagen at interfaces. Competitive adsorption against albumin and fibrinogen', *J. Biomed. Mater. Res.*, 21, 321-328, 1987.
3. Deyme, M., Ph.D. Thesis, Université Paris-Sud, 1989.
4. Hogland, P.D., 'Acylated β-caseins. Effect of alkyl group size on calcium ion sensitivity and on aggregation', *Biochemistry* 7, 2542-46, 1968.
5. Jentoft, N., Dearborn, D.G., 'Labelling of proteins by reductive methylation using sodium cyanoborohydride', *J. Biol. Chem.* 254, 4359-65, 1979.
6. Van der Scheer, A., Feijen, J., Klein Elhorst, J., Krugers Dagneaux, P.G.L.C. and Smolders C.A., 'The feasibility of radiolabel for Human Serum Albumin (HSA) adsorption studies', *J. Colloid and Interface Sc.* 66, 136-145, 1978.
7. Lyman, D.J., 'Structural order and blood compatibility of polymer prostheses', in: UPAC 'Structural order in polymers', F. Giardelli and P. Giusti (Eds.), Pergamon Press Oxford and New York, 205-220, 1981.
8. Van Oss, C.Y., Chandhury, M.K. and Good, P.J., 'Interfacial Lifshitz-van der Waals and polar interactions in macroscopic systems', *Chem. Rev. (American Chemical Society)* 88, 927-941, 1988.
9. Fukumura, H., Hayashi, K. and Yoshikawa, S., Proceedings of the Third Biomaterials Congress, Kyoto, April 22-25, 1988.

3. Protein adsorption and the interaction of human endothelial cells with polymer surfaces

A. DEKKER and T. BEUGELING

University of Twente, Department of Chemical Engineering, Enschede, The Netherlands

Introduction

A systematic study of the effects of polymer surface properties on the interaction with human endothelial cells (HEC) may lead to the development of improved small-diameter vascular grafts. HEC, suspended in culture medium containing 20% human serum adhere and spread on moderately water-wettable polymers (contact angle of about 40 degrees, Ref. 1). Examples of such polymers are: tissue culture polystyrene (TCPS), tissue culture poly(ethylene terephthalate) (TCPETP), cellulose-3-acetate, and a copolymer of hydroxyethyl methacrylate (HEMA) and methyl methacrylate (MMA), mol ratio 25 HEMA/75 MMA.

Reduced or no adhesion of HEC occurs on more hydrophilic materials such as poly (HEMA) and cellulose (Cuprophane), and on more hydrophobic materials such as poly(ethylene terephthalate) (PETP, Dacron), fluoroethylenepropylene copolymer (FEP, a Teflon-like material, Ref. 2) and Teflon (see below).

It was shown that, besides wettability, surface charge is an important property of a material with respect to HEC adhesion and spreading. For instance HEC adhere very well onto the very wettable, but positively charged, copolymer of HEMA and trimethylaminoethyl methacrylate (TMAEMA-Cl), mol ratio 85 HEMA/15 TMAEMA-Cl (Ref. 2, 3).

Polymers which show a good adhesion, spreading and proliferation of HEC, allow the cells to deposit (a) spreading factor(s) such as fibronectin, on the surface. This was demonstrated for TCPS. The use of culture medium containing fibronectin-depleted serum and the application of an anzyme-immunoassay in which a monoclonal anti-fibronectin antibody was used, proved that cellular fibronectin is deposited on TCPS during adhesion and spreading of HEC (Ref. 4, 5).

Y.F. Missirlis and W. Lemm (eds), Modern Aspects of Protein Adsorption on Biomaterials, 25-28.
© 1991 *Kluwer Academic Publishers. Printed in the Netherlands.*

Experiments and discussion

Earlier, we could not demonstrate that an exchange of preadsorbed protein for cellular fibronectin occurs (Ref. 4). Probably, a desorption of these proteins (albumin, high density lipoprotein and immunoglobulin G) could not be detected by the applied enzyme-immunoassay because a protein will generally show an adsorption plateau which is determined by a packed monolayer of antibody molecules.

However, after precoating of the TCPS by serum solutions in which 1% or less human serum was present, we could measure a decrease of the amount of adsorbed albumin after adhesion of endothelial cells and this decrease was accompanied by an increase of cellular fibronectin on the polymer surface (Fig. 1). This result indicates that an exchange of adsorbed proteins for cellular fibronectin is a prerequisite for adhesion and spreading of endothelial cells on a material surface.

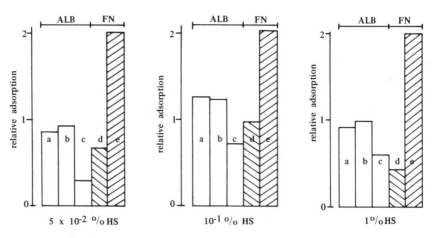

Fig. 1. Exchange of albumin, adsorbed from diluted serum to TCPS, for fibronectin deposited by adhering and spreading human endothelial cells.
TCPS was precoated with solutions of human serum (% HS) for 1 hour. Adsorption of albumin (ALB) from these solutions is indicated with *a*. After exposure of the precoated TCPS to serum-containing culture medium (which contains 20% human serum) for 6 hours, the amount of adsorbed albumin is *b*. The amount of fibronectin (FN) adsorbed from the culture medium is indicated with *d*. The total amount of fibronectin (FN), deposited after exposure (6 hours) of the precoated surface to a suspension of endothelial cells (20 • 10^4 cm^{-2}) in culture medium, is indicated with *e*. The amount of albumin, remaining on the TCPS surface, after exposure of the precoated surface to the cell suspension, is indicated with *c*.
Detection: enzyme-immunoassay.

Recently, we treated Teflon films (films of poly(tetrafluoroethylene), Fluorplast Nederland) with 'gas plasma' (because of the word plasma, we will use the expression: surface treatment or surface treated Teflon). Surface treatment was carried out in the presence of N_2 or O_2, which is indicated with N and O, respectively, in the Fig. 2 and 3.

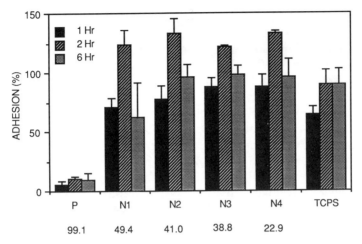

Fig. 2. Adhesion of human endothelial cells (6 • 10⁴ cm⁻²) to untreated Teflon (P) and surface treated Teflon (N_1-N_4). Cell adhesion to TCPS is also shown. The reference surface was TCPS, precoated with (crude) fribonectin (100% cell adhesion). Contact angles of the surfaces are also given. The contact angle of TCPS is 35° (captive bubble method in water).

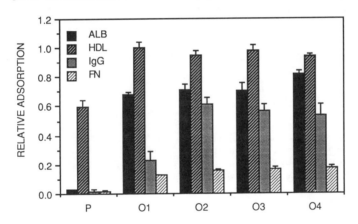

Fig. 3. Adsorption of albumin (ALB), high density lipoprotein (HDL), immunoglobulin G (IgG) and fibronectin (FN) from serum-containing culture medium (20% human serum) to untreated Teflon (P) and surface treated Teflon (O_1-O_4). Exposure time 1 hour. Detection: enzyme-immunoassay.
Contact angles of O_1, O_2, O_3 and O_4 are 50°, 44°, 28° and 17°, respectively (captive bublle method in water).

ESCA measurements revealed that surface treatment of Teflon in both O_2 and N_2 results, among other groups, in the formation of C-O-C, C = O and O-C = O groups. The fact that O was also present in N_2-treated surfaces while N was present in O_2-treated surfaces, may be explained by the formation of polymer-radicals which reacted with O_2 and N_2 from the air after the surface-treatment. Depending on the exposure time, surface-treated Teflon showed water-contact angles of 17° to 58°, while untreated Teflon shows a contact angle of 99°.

28

The adhesion of cultured human endothelial cells (HEC) from serum-containing culture medium to surface treated Teflon with contact angles of 20° to 45° leads to the formation of a monolayer of cells which is similar to the one on fibronectin-precoated TCPS (reference material). This is not the case on untreated Teflon. The adhesion of HEC to untreated and to surface-treated Teflon surfaces is shown in Fig. 2.

Expanded Teflon (ePTFE-patch, soft tissue 1310015020, Gore-Tex), which has been surface-treated, also shows an optimal adhesion of HEC.

The amount of various proteins adsorbed from the serum-containing culture medium to surface-treated Teflon seem to be much larger than the amount adsorbed to untreated Teflon (Fig. 3). Most probably, surface-treated Teflon not only shows a relatively high affinity for fibronectin, but also allows the displacement of previously adsorbed serum proteins for cellular fibronectin. Such a displacement will presumably occur relatively easy because surface treated Teflon is less hydrophobic than untreated Teflon.

Hydrophilic polymers like cellulose (Cuprophane) show reversible protein adsorption. This accounts for the findings that HEC do not adhere to Cuprophane and that precoating of this polymer with fibronectin (in order to improve cell adhesion) was found to be not effective (Ref. 2). The fact that surface-treated Teflon with relatively low contact angles ($\pm 20°$) adsorb large amounts of protein (Fig. 3) and show a good adhesion of HEC, is presumably due to the presence of charged groups on the surface.

References

1. Van Wachem, P.B., Beugeling, T., Feijen, J., Bantjes, A., Detmers, J.P., Van Aken, W.G., 'Interactions of cultured human endothelial cells with polymeric surfaces of different wettabilities', *Biomaterials* 1985, 6, 403-408.
2. Van Wachem, P.B., 'Interactions of cultured human endothelial cells with polymeric surfaces', Ph.D. Thesis, University of Twente, 1987.
3. Van Wachem, P.B., Hogt, A.H., Beugeling, T., Feijen, J., Bantjes, A., Detmers, J.P., Van Aken, W.G., 'Adhesion of cultured human endothelial cells onto methacrylate polymers with varying surface wettability and charge', *Biomaterials* 1987, 8, 323-328.
4. Van Wachem, P.B., Mallens, B.W.L., Dekker, A., Beugeling, T., Feijen, J., Bantjes, A., Detmers, J.P., Van Aken, W.G., 'Adsorption of fibronectin derived from serum and from human endothelial cells onto tissue culture polystyrene', *J. Biomed. Mater. Res.* 1987, 21, 1317-1327.
5. Van Wachem, P.B., Beugeling, T., Mallens, B.W.L., Dekker, A., Feijen, J., Bantjes, A., Van Aken, W.G., 'Deposition of endothelial fibronectin on polymeric surfaces', *Biomaterials* 1988, 9, 121-123.

4. Protein adsorption in relation to platelet adhesion and contact activation

A.A. POOT and T. BEUGELING

University of Twente, Department of Chemical Engineering, Enschede, The Netherlands

Introduction

The first event which takes place after contact of blood (or plasma) with an artificial surface is the rapid adsorption of proteins from the blood onto the material surface. It is generally assumed that all further events, such as platelet adhesion and surface activation of blood coagulation (contact activation), are determined by the composition and structure of the initially adsorbed protein layer.

From *in vitro* experiments it has long been realized that the adhesion of platelets is promoted when fibrinogen has been adsorbed to a material surface (Ref. 1, 2) and that platelet adhesion is reduced when preadsorbed albumin is present on the surface (Ref. 3, 4).

Lambrecht et al. (Ref. 5) studied the effect of preadsorption of several canine plasma proteins on surface-induced thrombogenesis in a canive *ex vivo* model. They found that plasticized poly(vinyl chloride) (PVC) precoated with von Willebrand factor (vWF, a protein synthesized by vascular endothelial cells), partially purified fibrinogen or fibronectin intensified platelet deposition compared with the uncoated PVC. The highest platelet deposition levels were found for preadsorbed vWF and partially purified fibrinogen.

The degree of contact activation is also dependent on the adsorbed protein layer, because the primary event of this process is an adsorption of clotting factor XII and a subsequent conformational change of this protein which is accompanied by the development of enzymatic activity against the plasma protein prekallikrein (Ref. 6). Prekallikrein circulates in plasma as a complex with high molecular weight kininogen (HMW kininogen) which accelerates the enzymatic conversion of prekallikrein into kallikrein. Kallikrein is the major plasma protease capable of cleaving factor XII. It must be noted that kallikrein cleaves surface-bound factor XII many times faster than fluid-phase factor XII. This is the second event whereby the foreign surface influences the contact activation.

The so called Vroman effect, which is characterized by a decrease of the amount of initially adsorbed fibrinogen from plasma onto glass-like surfaces

Y.F. Missirlis and W. Lemm (eds), Modern Aspects of Protein Adsorption on Biomaterials, 29-38.
© 1991 *Kluwer Academic Publishers. Printed in the Netherlands.*

with increasing contact time, has been attributed to a displacement of adsorbed fibrinogen by HMW kininogen (Ref. 7). Evidence was presented by Schmaier et al. (Ref. 7) and Brash et al. (Ref. 8) that HMW kininogen which has been cleaved by kallikrein (or activated factor XII) after contact activation, instead of intact HMW kininogen, is responsible for displacing adsorbed fibrinogen. Thus, the displacement of adsorbed fibrinogen from glass-like surfaces, and presumably other surfaces, may be correlated with the surface activation of blood coagulation. It is not likely that the amount of initially adsorbed fibrinogen itself influences the surface activation.

Platelet adhesion experiments

In our laboratory we used an *in vitro* capillary perfusion system according to Cazenave (Ref. 9, 10) for studying the effect of preadsorbed proteins on platelet deposition. In principle this system consists of a capillary tube connected to a syringe which is filled with a suspension of [111]Indium oxine labelled platelets and (unlabelled) erythrocytes (Fig. 1).

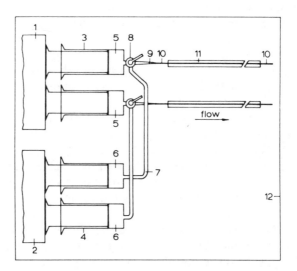

Fig. 1. The *in vitro* perfusion system according to Cazenave and Poot. 1 and 2, drive carriages of the two syringe pumps; 3 and 4, polypropylene syringes (20 or 50 ml); 5, protein solution or platelet suspension; 6, rinsing buffer; 7, PVC tube (3 mm ID, 25 cm long); 8, polycarbonate 3-way stopcock; 9, polypropylene pipette tip; 10, polyethylene capillary tube (0.75 mm ID, 30 cm long); 11, glass tube (1.5 mm ID, 25 cm long); 12, incubator 37°C.

Several precautions were made in order to warrant the reliability of the experimental results. Platelets and erythrocytes were isolated and washed according to Cazenave and Poot (Ref. 10). In this procedure PGI_2 is added during the steps in which platelets are washed to prevent platelet activation. The

2final buffer in which the cells are suspended contains Ca^{2+}, Mg^{2+} and apyrase. Apyrase is added in order to convert extracellular ADP into AMP. As a result platelet sensitivity to ADP-induced aggregation is restored and maintained. The suspensions of platelets and erythrocytes were prepared shortly before perfusion experiments and prewarmed for 5 min at 37°C. Experiments were also carried out at 37°C.

Medical grade high pressure polyethylene tubes, which were used in the perfusion system, were extensively rinsed with ethanol and distilled water respectively. The tubes were filled with Tyrode solution before introducing a protein solution or plasma for precoating the inner walls of the tubes with protein. In this way, air-liquid-solid interfaces were avoided. Such an interface was also avoided when cell suspensions were introduced into the capillary tubes.

In Fig. 2 the effect of protein precoating on platelet deposition is shown. Polyethylene tubes precoated with purified von Willebrand factor, fibrinogen or fibronectin strongly stimulated platelet deposition compared with the uncoated surface. Von Willebrand factor was the most adhesive protein, which is in agreement with the above mentioned results of Lambrecht et al. (Ref. 5), who used an *ex vivo* canine model. These results suggest that platelet adhesion is stimulated when an artificial surface adsorbs vWF, fibrinogen or fibronectin from blood.

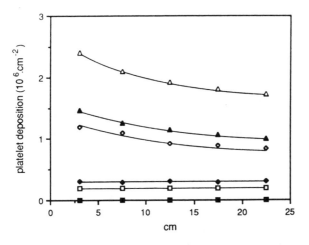

Fig. 2. Effect of protein precoating on platelet deposition. Polyethylene capillary tubes were precoated with Δ, von Willebrand factor; ▲, fibrinogen; ◇, fibronectin; □, IgG; ■, albumin and plasma; ◆, uncoated. Perfusions were carried out for 5 min at a shear rate of 500 s⁻¹. All values are the mean of 2 determinations.

Incubation of platelets with a monoclonal antibody which is directed against the platelet membrane complex GPIIb/IIIa and which inhibits both binding of fibrinogen to platelets in solution and platelet aggregation (Ref. 10) was found to completely inhibit platelet deposition to the inner walls of the tubes precoated with fibrinogen. Platelet deposition to polyethylene precoated with fibronectin or von Willebrand factor was inhibited to a certain extent by incubation of platelets with the monoclonal antibody. These results show that the GPIIb/IIIa complex plays an important role in platelet adhesion onto (pre)adsorbed proteins.

Scanning electron microscopy of the inner surfaces of the tubes revealed that few large platelet aggregates were present on the IgG precoated polyethylene. However, the number of deposited platelets per surface area was lower compared with uncoated surfaces. This was rather surprising because platelets possess receptors which can interact with (altered) IgG. Probably, platelet aggregates had been detached from the surface due to shear forces.

Platelet adhesion was almost completely absent on polyethylene precoated with albumin or plasma. Obviously, precoating with plasma for one hour at 37°C generates a protein layer which is unfavourable for platelet adhesion. It has been shown that the hydrophobic polymers PVC and polystyrene (Ref. 11) as well as polyethylene (Ref. 10) show a preferential adsorption of high density lipoprotein (HDL) from plasma (see below). Recently, we found that platelet deposition is virtually absent on polyethylene precoated with HDL. Therefore, the very few platelets adhering to polyethylene precoated with plasma may be due to a relatively large amount of HDL present on the polyethylene surface.

Protein adsorption experiments

Protein adsorption from plasma to polyethylene and glass was studied by means of a two step enzyme-immunoassay (Ref. 10). A limitation of this technique is that the detection of adsorbed (lipo)proteins is only semi-quantitative. Therefore, one may not take a quantitative comparison between 'measured' amounts of different proteins, adsorbed to a material surface.

In Fig. 3 the test device for the adsorption experiments is shown. In order to prevent the formation of an air-liquid-solid interface, the wells were partially filled with PBS buffer before introducing plasma (or a plasma solution). For this reason the highest plasma concentration in our experiments was 50% (v/v).

In Figs. 4 and 5 the adsorption of fibrinogen, HDL and HMW kininogen from 1:1 diluted plasma to polyethylene and glass as a function of time is shown. The amount of initially adsorbed HMW kininogen on polyethylene decreased during the first few minutes of contact time; thereafter a small amount of adsorbed HMW kininogen was detected on the surface. The relatively small amount of adsorbed HDL which was detected on polyethylene is most probably caused by the fact that only part of the HDL molecule consist of protein apo A1. As a consequence the surface concentration of antibodies directed against apo A1 is much lower compared to the surface concentration of antibodies in case of a pure protein adsorbed to the surface.

It must be mentioned that maxima are found for the adsorption of fibrinogen, HMW kininogen, IgG, fibronectin and albumin to polyethylene as

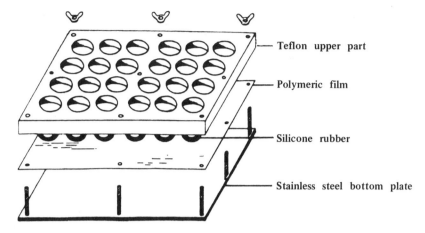

Fig. 3. Test chamber. The Teflon upper part is provided with 24 cyclindrical holes (10.0 mm ID). Each hole is provided with a stepped recess at the bottom side in which a sealing ring of silicone rubber is placed.

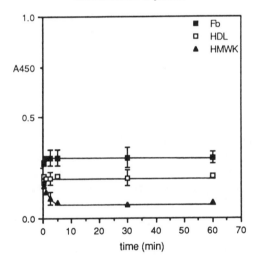

Fig. 4. Adsorption of fibrinogen (Fb), HMW kininogen (HMWK) and HDL from 1:1 diluted human pooled plasma to polyethylene as a function of time. All values are the mean of 4 determinations.

a function of plasma concentration, whereas HDL shows an adsorption plateau at higher plasma concentrations (Fig. 6). Similar results have also been reported for the adsorption of HDL from plasma to PVC and polystyrene (Ref. 11). Evidence was presented that these polymers show a preferential adsorption of HDL. It may be concluded that polyethylene also shows a preferential adsorption of HDL from plasma.

The amounts of fibrinogen and HDL initially adsorbed from 1:1 diluted plasma onto glass decreased during the first five minutes of contact time.

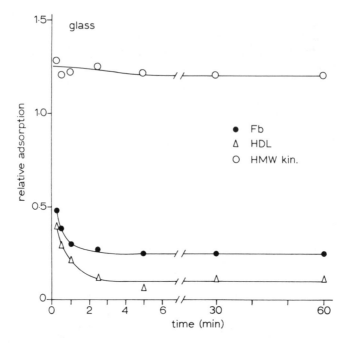

Fig. 5. Adsorption of fibrinogen (Fb), HMW kininogen (HMW kin.) and HDL from 1:1 diluted human pooled plasma to glass as a function of time. All values are the mean of 4 determinations.

Hereafter very small amounts of these plasma components seem to be adsorbed to the glass surface. The amount of HMW kininogen adsorbed to glass (almost) instantly reached a high value. Moreover HMW kininogen, adsorbed from plasma solutions to glass, shows an adsorption plateau above a plasma concentration of 1%, whereas the amounts of fibrinogen HDL, IgG, fibronectin and albumin decrease at plasma concentrations above 0.1 – 1%. Evidently, HMW kininogen is preferentially adsorbed to glass.

Fig. 5 shows that only a small amount of adsorbed fibrinogen was detected on glass after about 5 minutes of exposure to 1:1 diluted plasma. The 'desorption' of fibrinogen from glass was markedly less when 1:1 diluted HMW kininogen-deficient plasma was used (Fig. 7).

Our results roughly agree with the results of Vroman et al. (Ref. 12), Schmaier et al. (Ref. 7) and Brash et al. (Ref. 8) and seem to support the hypothesis of Vroman et al. (Ref. 12) that fibrinogen, initially adsorbed from plasma onto hydrophilic surfaces like glass, is displaced by HMW kininogen. However, the fact that the amount of adsorbed HMW kininogen does not seem to increase during the first few minutes while the amount of initially adsorbed fibrinogen decreases, suggests that HMW kininogen is probably not the only protein which is responsible for the displacement of fibrinogen adsorbed from plasma to glass-like surfaces.

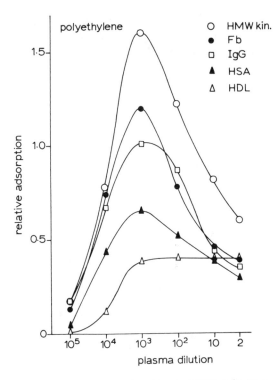

Fig. 6. The adsorption of the human proteins albumin (HSA), IgG, fibrinogen (Fb), HMW kininogen (HMW kin.) and HDL to polyethylene as a function of plasma concentration.

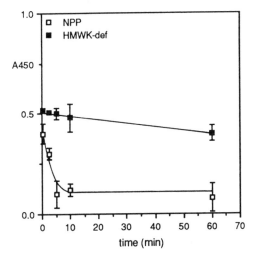

Fig. 7. Adsorption of fibrinogen from 1:1 diluted normal pooled plasma (NPP) and HMW kininogen-deficient plasma (HMWK-def) onto glass as a function of time. All values are the mean of 4 determinations.

Fig. 8 shows the adsorption of fibrinogen and single-chain HMW kininogen from solutions of a binary mixture of these components to glass. From this figure it may be concluded that the adsorption of fibrinogen is not affected by the presence of single-chain HMW kininogen. This is not surprising if it is supposed that HMW kininogen, cleaved by kallikrein or activated factor XII, is preferentially adsorbed to glass or may be responsible for a displacement of fibrinogen from glass (see Introduction).

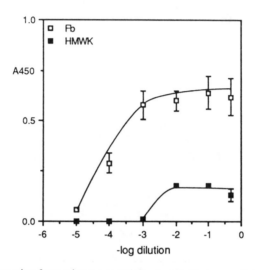

Fig. 8. Protein adsorption from mixtures containing human fibrinogen (3 mg/ml) and human HMW kininogen (70 μg/ml) onto glass. Exposure time 1 hour. All values are the mean of 4 determinations.

Conclusions

From the results presented here it may be concluded that artificial surfaces which adsorb von Willebrand factor, fibrinogen or fibronectin from blood (or plasma), show an enhanced platelet adhesion and as a result thrombus formation.

On hydrophobic polymers, the small amounts of adsorbed fibrinogen and other proteins may be due to a preferential adsorption of HDL. The presence of HDL on such polymers probably causes a passivation of the surfaces resulting in a reduced platelet adhesion.

Our results suggest that one or more plasma proteins, besides HMW kininogen, are involved in the displacement of fibrinogen which initially adsorbs from plasma onto glass-like surfaces.

References

1. Packham, M.A., Evans, G., Glynn, M.F. and Mustard, J.F. (1969), 'The effect of plasma proteins on the interaction of platelets with glass surfaces', *J. Lab. Clin. Med.* 73, 686-697.
2. Zucker, M.B. and Vroman, L. (1969), 'Platelet adhesion induced by fibrinogen adsorbed onto glass', *Proc. Exp. Soc. Biol. Med.* 131, 318-320.
3. Lee, R.G. and Kim, S.W. (1979), 'Adsorbed glycoproteins in platelet adhesion onto polymer surfaces: significance of terminal galactose units', *Trans. Amer. Soc. Artif. Intern. Organs* 25, 124-131.
4. Olijslager, J. (1982), 'The development of test devices for the study of blood material interactions', Ph.D. Thesis, Twente University, Enschede, The Netherlands.
5. Lambrecht, L.K., Young, B.R., Stafford, R.E., Park, K., Albrecht, R.M., Mosher, D.F. and Cooper, S.L. (1986), 'The influence of preadsorbed canine von Willebrand factor, fibronectin and fibrinogen on *ex vivo* artificial surface-induced thrombosis', *Thromb. Res.* 41, 99-117.
6. Kaplan, A.P. and Silverberg, M. (1987), 'The coagulation-kinin pathway of human plasma', *Blood* 70, 1-15.
7. Schmaier, A.H., Silver, L., Adams, A.L., Fischer, G.C., Munoz, P.C., Vroman, L. and Colman, W. (1983), 'The effect of high moleculer weight kininogen on surface-adsorbed fibrinogen', *Thromb. Res.* 33, 51-67.
8. Brash, J.L., Scott, C.F., ten Hove, P., Wojciechowski, P. and Colman, W. (1988), 'Mechanism of transient adsorption of fibrinogen from plasma to solid surfaces: role of the contact and fibrinolytic systems', *Blood* 71, 932-939.
9. Cazenave, J.P., Mulvihill, J.N., Huisman, J.G., and Van Aken, W.G. (1987), 'The use of monoclonal antibodies against platelet membrane glycoproteins to measure platelet accumulation on artificial surfaces', in: *Biology and Pathology of the Platelet: Vessel Wall Interactions*, Jolles, G., Legrand, Y. and Nurden, A. (Eds.), Academic Press, New York, 1987, p.375.
10. Poot, A.A. (1989), 'Protein adsorption and platelet deposition on biomaterials: *in vitro* studies concerning blood compatibility', Ph.D. Thesis, Twente University, Enschede, The Netherlands.
11. Breemhaar, W., Brinkman, E., Ellens, D.J., Beugeling, T. and Bantjes, A. (1984), 'Preferential adsorption of high density lipoprotein from blood plasma onto biomaterial surfaces', *Biomaterials* 5, 269-274.
12. Vroman, L., Adams, A.L., Fischer, G.C. and Munoz, P.C. (1980), 'Interaction of high molecular weight kininogen, factor XII and fibrinogen in plasma at interfaces', *Blood* 55, 156-159.
13. Poot, A., Beugeling, T., Bantjes, A. and Van Aken, W.G., 'Platelet deposition in a capillary perfusion model: quantitative and morphological aspects', *Biomaterials* 1988, 9, 126-132.
14. Dekker, A., Poot, A., Beugeling, T., Bantjes, A.and Van Aken, W.G., 'The effect of vascular cell seeding on platelet deposition in an *in vitro* capillary perfusion model', *Thromb. Haemostas.* 1989, 61(3), 402-408.
15. Riethorst, W., 'Affinity chromatography for large scale purification of coagulation factor VIII', Ph.D. Thesis, University of Twente, Enschede, The Netherlands, 1988.
16. Heuvelsland, W.J.M., Dubbeldam, G.C. and Brouwer, W., 'Reflectometry as a tool for dynamically measuring protein adsorption', in: S. Dawids (Ed), *Test procedures for blood compatibility of biomaterials*, to be published.
17. Welin, S., Elwing, H., Arwin, H., Lundstrom, I. and Wikstrom, M., 'Reflectometry in kinetic studies of immunological and enzymatic reactions on solid surfaces', *Anal. Chem. Acta* 1984, 163, 263-267.
18. Brinkman, E., Poot, A., Beugeling, T., Van der Does, L. and Bantjes, A., 'Surface modification of copolyether-urethane catheters with poly(ethylene oxide)', *Int. J. Art. Organs* 1989, 12, 390-394.
19. Aalto, S., Manuel, H.J., Van der Does, L. and Bantjes, A., 'Poly(N-vinylacetamide) and

poly(N-methyl-N-vinylacetamide) hydrogels for biomedical applications', Macromolecular preprints '89 (2nd Euro-American Conference in Oxford on Functional Polymers and Biopolymers, 4-8 September, 1989), p. 55.

5. Studies of protein adsorption relevant to blood compatible materials

J.L. BRASH

Departments of Chemical Engineering and Pathology, McMaster University, Hamilton, Ontario, Canada

Introduction

The ultimate objective of the work in our laboratory is to develop blood compatible materials. To this end two principal lines of research are currently being pursued:
1. Materials development per se, with current effort concentrated on polyurethanes. This work will not be discussed in the present paper.
2. The study of blood-material interactions, the main focus of the present discussion. We do this type of work on the basis that rational design of biomaterials does not seem possible in an information 'vacuum'. We need to have the information required to understand the mechanisms of coagulation, thrombus formation and other blood-surface interactions.

Most of the work in our laboratory in this area is concerned with protein adsorption. Our early work on this problem dealt with simple systems of one or two proteins. Results from this work are summarized below. Our current work represents a move towards what might be seen as the 'real world' of plasma and blood. In this context the answers to three questions are being sought:
1. What is the composition of the protein layer which is laid down on different surfaces in contact with blood and what is the variation of that composition with the type of surface?
2. How does the composition of the protein layer change as a function of time?
3. What are the 'consequences' of protein adsorption, i.e. what are the secondary events that take place? Does the protein simply 'sit' on the surface, i.e., is it simply bound to the surface, or does it bind and then undergo different types of transformation on the surface: denaturation, changes of biological activity etc.?

It is hoped that all of this work will help to elucidate the principles required for developing biocompatible materials. In the present author's opinion, this may well reduce to the question: how we can control the composition of the protein layer which is laid down on a material from blood, from tissue, from tear fluid, or from any other biological medium with which it is placed in contact? The idea of controlling adsorption to achieve a given protein layer

Y.F. Missirlis and W. Lemm (eds), Modern Aspects of Protein Adsorption on Biomaterials, 39-47.
© 1991 *Kluwer Academic Publishers. Printed in the Netherlands.*

composition then emerges as a strategy for designing biocompatible interfaces.

Single protein studies

The results of our studies on single protein adsorption may be summarized as follows (Brash, 1987).

First, protein adsorption is ubiquitous: it is axiomatic that when protein solutions are put in contact with solid surfaces adsorption occurs.

Second, adsorption is rapid (Chan and Brash, 1981a). Adsorption kinetics is difficult to discuss in general terms, because it is physical system dependent. It depends on flow, agitation conditions, whether one is working with tubing, sheet, particle, etc. However it is generally accepted that surfaces are substantially covered within a few seconds, or at most, a few minutes of contact.

Third, monolayers of proteins are generally formed (Chan and Brash, 1981a; Schmitt et al., 1983) and multilayer formation is rare. The density of the monolayer varies with the surface type. Sometimes dense monolayers, and sometimes relatively sparse monolayers are formed. Density tends to increase with surface hydrophobicity.

Fourth, adsorption is inherently reversible. This has been an area of dispute ever since protein adsorption began to be studied. Given the present state of knowledge, the adsorption of proteins appears to this author to be inherently, but only slowly reversible. However, informed opinion is by no means unanimous on this point. Two pieces of evidence argue convincingly that protein adsorption is reversible. First, the initial slopes of protein adsorption isotherms are finite as opposed to infinite (Schmitt et al., 1983). Second, self exchange has been shown to occur in many systems. Thus if one does a single protein adsorption experiment and maintains the protein solution in contact with the surface, one can demonstrate that there is exchange between the solution and the surface (Brash and Samak, 1978; Brash et al., 1983). Such behaviour reflects an inherent reversibility. If the same phenomena occur in a complex multicomponent system then a mechanism exists for the composition of the layer to change as a function of time.

Fifth, denaturation of adsorbed proteins may occur although one cannot easily generalize in this regard. Each protein-surface system must be considered individually. We have shown in this connection that fibrinogen eluted from glass suffers a loss of α-helicity (Chan and Brash, 1981b) and sometimes it is degraded in a pattern suggesting that there are enzymes in the sample which cause specific degradation of the adsorbed protein (Brash et al., 1985).

We have also shown, using a hydrodynamic technique, that the thickness of an adsorbed layer of a single protein can change if one changes the ambient conditions. For example cycling the pH of the solution in contact with fibrinogen adsorbed on glass causes the layer thickness to increase and then decrease (de Baillou et al., 1984). Since there is no change in the quantity of protein in the layer it appears that the protein extends away from the surface and then retracts as the pH is increased and then decreased.

In general, single protein systems are well adapted to the study of denaturation using physical techniques. This has been a contribution of single protein studies (Morrissey and Stromberg, 1974; Hlady et al., 1986).

Plasma and blood studies

In more recent work, as previously mentioned, we have been studying the protein layer composition following material-blood contact. Most of our earlier work on single proteins was done using radioiodine labelled proteins. The experimental approach in our current work on adsorbed layer composition is simply to contact the material with plasma or blood, then elute the surface with different buffers. We then attempt to identify the proteins which are present in the eluted protein mixture. Three major points should be made from the data thus far (Brash and Thibodeau, 1986; Mulzer and Brash, 1989; Boisson-Vidal et al., 1991):

1. The mixtures which are eluted from different surfaces in contact with either plasma or blood are complex, multicomponent mixtures. Most proteins are adsorbed on most surfaces.
2. The main difference among surfaces in contact with plasma or blood is that the relative amounts of the different proteins vary from surface to surface.
3. We have found very recently that during whole blood contact adsorbed proteins are often degraded. This may be due to the presence of proteolytic enzymes derived from damaged cells which are able to degrade the proteins on the surface. This seems to be a new, or at least hitherto unremarked, finding.

With respect to these conclusions, Fig. 1 shows SDS-PAGE data for protein mixtures eluted from glass after contact with plasma (Brash and Thibodeau, 1986). Lane 1 shows proteins eluted by a high ionic strength buffer (1 M Tris), lane 2 shows proteins eluted by SDS, and lane 3 shows a fibrinogen sample for reference purposes. An important point is that there are many different protein bands in the eluate gels. If one does the same experiment on different surfaces one gets equally complex, but different patterns (Boisson-Vidal et al., 1991).

SDS-PAGE gives only an indication of molecular weight but does not provide unambiguous identification of proteins. Therefore we have recently been using immunoblotting methods which give positive identification of the eluted proteins (Gershoni and Palade, 1983). In one set of experiments we collected dialyzers after use by patients and eluted the proteins from the membranes. Following SDS-PAGE of the eluted proteins we blotted the gels onto a PVDF (Immobilon) membrane and then probed strips of the blots using antibodies to different proteins as indicated in Fig. 2 for a Cuprophan dialyzer (Mulzer and Brash, 1989).

42

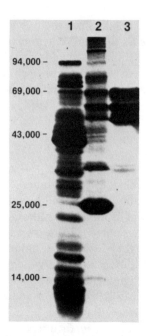

Fig. 1. SDS-PAGE of proteins eluted from glass after contact with normal human plasma: Lane 1: 1M tris eluate. Lane 2: SDS eluate. Lane 3: purified fibrinogen. (Reproduced with permission from Brash and Thibodeau, 1986).

We tested for sixteen proteins in all and as can be seen there are bands in various positions for all of them. Thus during dialysis all of these sixteen proteins are adsorbed to the membrane surfaces. A major conclusion again is therefore that the layer adsorbed during blood contact is complex. Also if one considers the high molecular weight proteins, for example complement C3 and α_2 – macroglobulin, they appear to be extensively degraded. The blot using a C3 antibody reveals an extremely complex pattern of bands indicating that adsorbed complement C3 is extensively degraded in a non-specific fashion on this dialyzer membrane. Normally one sees only two bands in a blot of complement C3: one at about 70 kD (the β-chain) and one at about 110 kD (the α-chain). We believe enzymes are being released from blood cells which degrade the proteins adsorbed on Cuprophan membranes. It should be pointed out that the complexity of the C3 blot precludes any conclusion on complement activation from these data.

Another approach we are pursuing at the present time in relation to blood and plasma is to study the temporal evolution of the protein layer, i.e. the manner in which it changes composition as a function of time. Our efforts have focused mainly on the Vroman effect (Vroman et al., 1980; Horbett, 1984; Brash and ten Hove, 1984). Although there is still an element of 'mystique' associated with this phenomenon we believe the Vroman effect tells us simply that the composition of the adsorbed layer from blood or plasma follows some

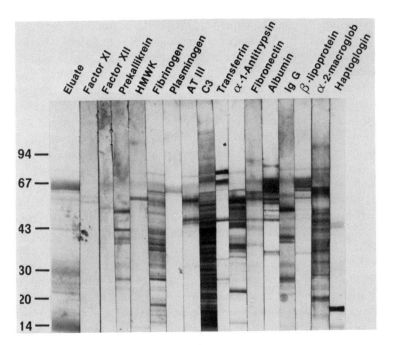

Fig. 2. Immunoblot from SDS-PAGE (reduced) of 2% SDS eluate from a Cuprophane dialyzer after clinical use. Left lane: Amido black-stained eluate sample. Remaining lanes: Immunostaining patterns for specific antisera as indicated. Molecular weight scale in kD. (Reproduced with permission from Mulzer and Brash, 1989).

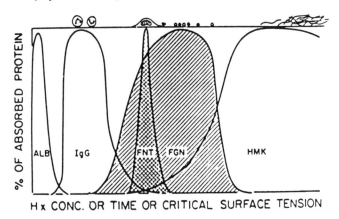

Fig. 3. Conditions affecting composition of proteins absorbed out of normal intact plasma. Relative concentrations of proteins expected to remain adsorbed under a choice of conditions shown along abscissa. (Reproduced with permission from Vroman, 1987).

kind of sequence. We may not yet know all the details of the sequence such as order, timing and surface type dependence, but it seems to be generally accepted that there is a sequence. Vroman has portrayed the mechanism as an adsorption sequence as shown in Fig. 3 (Vroman, 1987). In our opinion Fig. 3 should not be taken literally. Even if the sequence portrayed is correct with regard to which protein follows which, the Figure is somewhat misleading since it suggests that proteins come to the surface more or less discretely, that is one at a time. In fact it is evident that all proteins begin to be adsorbed immediately though at different rates dependent on diffusion. It is likely that they are then displaced sequentially depending on relative binding affinity. The overall effect is the same for both mechanisms, i.e. a sequence of layer composition.

There is now considerable evidence in the literature to support this kind of mechanism (Vroman, 1987; Vroman and Adams, 1986; Slack and Horbett, 1988.

The experiment used in our lab which shows the Vroman effect very nicely, involves adsorption of fibrinogen from plasma. Fig. 4 shows kinetic data for a glass surface and the different curves are for plasma diluted to different extents (Brash and ten Hove, 1984). For moderately diluted plasma there is an initial increase in fibrinogen adsorption followed by a decrease. As one dilutes the plasma the increase is slower, but there is still a maximum in adsorption as a function of time. If one dilutes the plasma sufficiently one gets 'conventional' kinetics with fibrinogen adsorption simply increasing and then levelling off. The data are consistent with the Vroman model in that as one dilutes the plasma one reaches a point where there is insufficient of the replacing components to cause displacement of the initially adsorbed fibrinogen.

It is perhaps easier to consider this kind of data in the so-called 'concentration domain' as shown in Fig. 5 (Brash and ten Hove, 1989). This shows fibrinogen adsorption from plasma onto glass as a function of plasma concentration for a fixed time, the plasma being diluted with a buffer. For 5-

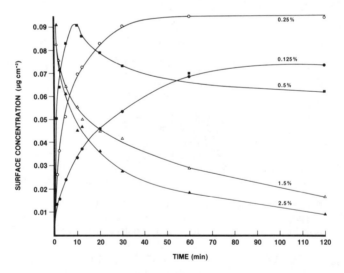

Fig. 4. Kinetics of adsorption of fibrinogen to glass from plasma at varying plasma concentrations. Plasma concentrations are indicated on curves. (Reproduced with permission from Brash and ten Hove, 1984).

minute exposures fibrinogen adsorption increases with plasma concentration, a peak occurs around 1% plasma, and then adsorption decreases. One can do this experiment in either plasma or whole blood as shown in Fig. 5. Not only are the data qualitatively the same, they are quantitatively almost identical. So this is a 'real world' effect; it does occur in whole blood.

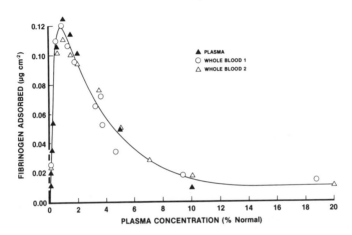

Fig. 5. Adsorption of fibrinogen from plasma or blood to glass at varying plasma concentrations. Plasma or blood diluted with isotonic tris buffer, pH 7.35. Adsorption time 5 min. (Reproduced with permission from Brash and ten Hove, 1989).

A final aspect of the Vroman effect which should be mentioned is that it is not seen on all surfaces. Figure 6 shows fibrinogen adsorption at a fixed time, 5 min, as a function of plasma concentration for several surfaces (Santerre et al, 1989). All of the surfaces are polyurethanes. Data are shown for Biomer, a commercial polyurethane, and two others (ED and MDA) that may be considered 'conventional' which were synthesized in our laboratory. In addition there are data for a polyurethane containing sulfonate groups in the hard segment (BDDS-4). This material was also synthesized in our laboratory. It should be noted that the y-axis scale in Figure 6 is greatly compressed compared to those in the previous Vroman effect Figures.

Biomer and the other conventional polyurethanes show normal Vroman effect behaviour. The sulfonated material shows extraordinarily high fibrinogen adsorption from plasma and the Vroman effect has been eliminated. The high level of adsorption on this surface may be partly due to the fact that it is strongly hydrophilic and takes up water (and perhaps protein) into the matrix of the material. We have shown in addition that this material does have some anticoagulant activity, and we may speculate that there may be a connection between its anticoagulant behaviour and the lack of a Vroman effect.

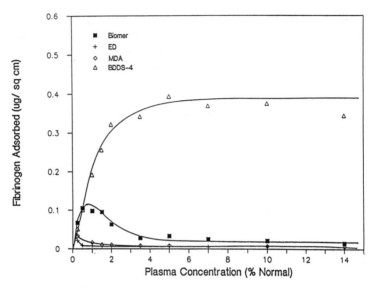

Fig. 6. Adsorption of fibrinogen from plasma as a function of plasma concentration to Biomer and experimental polyurethanes (5 minute exposure). ED and MDA are conventional polyurethanes. BDDS-4 is a polyurethane containing sulfonate groups in the hard segment.

Conclusions

Our work thus far on protein adsorption in relation to blood compatibility shows that the protein layers adsorbed from blood or plasma are complex, variable from surface to surface, and variable over time. Information on protein adsorption from blood is just beginning to appear in the literature from a number of laboratories. We are confident that this information will point the way to designing surfaces that respond to blood contact in a given manner, and in particular in a manner that minimizes the coagulation-thrombosis syndrome.

Q. Baszkin: In the slides you have shown to us you are counting adsorption of fibrinogen which is your labelled protein. Did you make the same experiments, in plural, with labeling other proteins which are of interest and following, if do they manifest and Vroman effect...

Answer: Yes we have. We've done it with IgG in particular and we've confirmed the Vroman effect in that case. In the case of albumin I think it's happening too fast. Maybe your technique would be better adapted for doing that.

This is not an in situ technique. I didn't take time to describe the method because it's been reported in the literature, already. But it is a non in situ technique. Thirty seconds is the minimum time at which you can make a measurement, because you have to displace the initial solution from the system, which takes a certain amount of time.

So, it's really a low time resolution experiment. We can detect the Vroman effect for IgG, but I think albumin is too rapid.

Acknowledgements

The author thanks his coworkers who have contributed to the work discussed in this paper: S. Uniyal, B.M.C. Chan, P. ten Hove, S. Mulzer, R. Cornelius, P. Wojciechowski, P.Santerre, N. Vanderkamp and K. Woodhouse. The financial support of the Medical Research Council of Canada, the Natural Sciences and Engineering Research Council of Canada, the Heart and Stroke Foundation of Ontario and the Ontario Centre for Materials Research is gratefully acknowledged.

References

1. Boisson-Vidal, C., Jozefonvicz, J. and Brash, J.L., *J. Biomed. Mater. Res.* 25, 67, 1991.
2. Brash, J.L., *Amer. Chem. Soc. Symp. Series* 343, 490, 1987.
3. Brash, J.L. and Samak, Q.and J. Colloid. *Interface Sci.* 65, 494, 1978.
4. Brash, J.L. and Ten Hove, P., *Thromb. Haemostas.* 51, 326, 1984.
5. Brash, J.L. and Ten Hove, P., *J. Biomed. Mater. Res.* 23, 157, 1989.
6. Brash, J.L. and Thibodeau, J.A., *J. Biomed. Mater. Res.* 20, 1263, 1986.
7. Brash, J.L. et al., *J. Colloid Interface Sci.* 95, 28, 1983.
8. Brash, J.L. et al., *J. Biomed. Mater. Res.* 19, 1017, 1985.
9. Chan, B.M.C. and Brash, J.L., *J. Colloid Interface Sci.* 82, 217, 1981a.
10. Chan, B.M.C. and Brash, J.L., *J. Colloid Interface Sci.* 84, 263, 1981b.
11. de Baillou, N. et al., *J. Colloid Interface Sci.* 100, 167, 1984.
12. Gershoni, J.M. and Palade, G.E., *Analyt. Biochem.* 131, 1, 1983.
13. Hlady, V., Reinecke, D.R. and Andrade, J.D., *J. Colloid Interface Sci.* 111, 555, 1986.
14. Horbett, T.A., *Thromb. Haemostas.* 51, 174, 1984.
15. Morrissey, B.W. and Stromberg, R.R., *J. Colloid Interface Sci.* 46, 152, 1974.
16. Mulzer, S.R. and Brash, J.L. *J. Biomed. Mater. Res.* 23, 1483, 1989.
17. Santerre, P. et al., *Trans. Soc. Biomaterials* 12, 113, 1989.
18. Schmitt, A. et al., *J. Colloid Interface Sci.* 92, 25, 1983.
19. Slack, S.M. and Horbett, T.A., *J. Colloid Interface Sci.* 124, 535, 1988.
20. Vroman, L., *Seminars Thromb. Haemostas.* 13, 79, 1987.
21. Vroman, L. and Adams, A.L., *J. Colloid Interface Sci.* 111, 391, 1986.
22. Vroman, L. et al., *Blood* 55, 156, 1980.

6. Methods to study blood-surface interactions

J.P. CAZENAVE and J.N. MULVIHILL

INSERM U.311, Biologie et Pharmacologie des Interactions du Sang avec les Vaisseaux et les Biomateriaux, Centre Regional de Transfusion Sanguine, Strasbourg, Cedex, France

Blood interaction with artificial surfaces is largely governed by the first step of the interaction process: adsorption of plasma proteins. Subsequent platelet and leukocyte adhesion and aggregation, activation of the coagulation, fibrinolysis and complement systems and thrombus formation are conditioned by the nature of the layer of adsorbed proteins, in particular its composition and conformational state (Ref. 1, 2).

Platelets are anucleated circulating cells produced from megakaryocytes in the bone marrow, normal non-activated platelets being discoid with a mean diameter of 2 μm and a mean volume of 5-7 μm^3 (Ref. 3). Activation of platelets by surface contact results in shape change, adhesion and spreading. Surface bound aggregates are generated by platelet-platelet interactions and concomitant activation of the coagulation system leads to formation of a platelet-fibrin thrombus.

As shown in Fig. 1, the platelet is a contractile and secretory cell whose structure can be divided into three major zones. The *peripheral membrane zone*, characterized by an asymmetric molecular distribution, is of particular importance for its content of (i) phospholipids, precursors of platelet agonists (arachidonic acid, platelet activating factor) and procoagulant platelet factor 3, and (ii) glycoproteins (GP), some of which form specific receptors for proteins involved in adhesion and aggregation (GP Ia-IIa for collagen, GP Ib-IX for von Willebrand factor (vWF), GP IIb-IIIa for fibrinogen, fibronectin and vWF, GP IV for thrombospondin). Covering the outer surface of the membrane is the glycocalyx, composed of adsorbed plasma proteins. The open canicular system, formed by invaginations of the plasma membrane, provides access to the interior cytoplasm and the exit route for secretion of the contents of storage granules, while the dense tubular system is the site of calcium storage and prostaglandin synthesis.

In the *cytoplasmic zone*, microtubules maintain the platelet discoid shape. Contractile proteins, microfilaments of actin and myosin, govern the consistency of the cytoplasm and control shape change from disc to sphere, pseudopod formation, granule secretion and clot retraction.

The *organelle zone* consists of mitochondria, glycogen, lysosomes and

Y.F. Missirlis and W. Lemm (eds), Modern Aspects of Protein Adsorption on Biomaterials, 49-53.
© 1991 *Kluwer Academic Publishers. Printed in the Netherlands.*

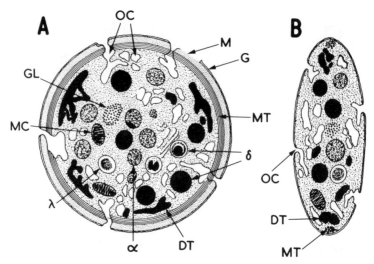

Fig. 1. Diagram of a platelet cut in equatorial (A) and longitudinal (B) planes. The plasma membrane (M) is covered by an exterior glycocalyx (G). Microtubules (MT) formed of tubulin lie in the cytoplasm immediately beneath in a circumferential band. The cytoplasm also contains glycogen (GL), and a number of organelles: mitochondria (MC); the (α) alpha and (δ) dense granules; lysosomes (λ); and a system of communication channels within the cell, the open canalicular system (OC) and the dense tubular system (DT).

storage granules. Dense granules contain ADP, ATP, serotonin and ionic calcium, while α-granules contain adhesive proteins (vWF, fibrinogen, fibronectin, thrombospondin), coagulation factor V and platelet specific proteins (β-thromboglobulin, platelet factor 4, platelet derived growth factor), products which are secreted after platelet stimulation by agonists such as collagen or thrombin or by contact with a foreign surface.

In our laboratory, we have developed a capillary perfusion system enabling quantitative evaluation of protein adsorption, platelet activation and accumulation and activation of coagulation and complement components in the course of blood interaction with biomaterials in catheter form (Ref. 4). Blood flow, controlled by a syringe pump, is non-pulsatile and laminar of Poiseuille type (Fig. 2). Air-liquid interfaces are strictly avoided and wall shear rates (0-4,000 s^{-1}) cover the range of interest for the cardiovascular system. Protein adsorption is determined using ^{125}I-labeled purified proteins, while platelet accumulation is measured either by ^{111}In-labeling of washed human platelets or by surface phase radioimmunoassay with a ^{125}I-labeled monoclonal antibody directed specifically against the platelet membrane glycoprotein IIb-IIIa.

Morphological examination of the surface by scanning electron microscopy allows distinction between platelet adhesion and formation of surface bound aggregates. Activation of platelets and the coagulation and complement systems may be quantitated by assay of platelet release products (β-thromboglobulin, platelet factor 4), fibrinopeptide A (FPA) and complement fragment C_{3a} in the influent and effluent blood.

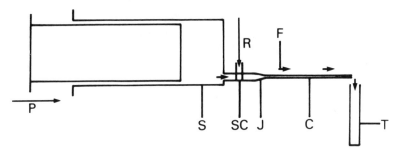

Fig. 2. Schematic diagram of the in vitro capillary perfusion system. S: 50 ml plastic syringe containing protein or antibody solution, anticoagulated whole blood or washed platelet suspension; P: piston driven by syringe pump; SC: 3-way stop-cock; J: joint in silicone tubing; C: glass capillary (0.80 or 0.56 mm i.d.); F: direction of blood flow; T: plastic tube for blood collection; R: rinsing buffer from a reservoir at 37°C, flow controlled by a peristaltic pump. The entire apparatus is enclosed in a thermostated hood at 37°C.

Initial experiments using protein coated glass capillaries perfused with washed platelet suspensions or anticoagulated whole blood demonstrated clearly the influence of the preadsorbed protein layer on platelet accumulation, especially the passivating effect of albumin as compared to the activating effect of fibrinogen or collagen (Ref. 4, Table 1). Passivation of artificial surfaces by albumin preadsorption has in fact proved to be of considerable clinical value. In therapeutic plasmapheresis, preadsorption of the extracorporeal circuit with 4% human serum albumin reduced the incidence of circuit blockage by platelet-fibrin thrombi from 44% to 4.6% of apheresis sessions (Ref. 5).

In the presence of adhesive proteins such as vWF, platelet deposition in the capillary system is enhanced. For instance, on albumin coated glass capillaries perfused with a washed platelet suspension for 2 min at 2,000 s^{-1}, the surface concentration of adherent platelets increased by a factor of 1.87 as compared to control values when vWF (5 μg/ml) was added to the suspension medium and by a factor of 4 when vWF was preadsorbed onto the surface (Ref. 6). Similar experiments have shown the potential importance of thrombin generation during blood contact with artificial surfaces in promoting platelet accumulation and thrombus formation (Ref. 7). Using protein coated glass capillaries perfused with a washed platelet suspension for 2 min at 2,000 s^{-1}, platelet deposition was found to increase by up to six-fold in the presence of surface adsorbed thrombin (~ 0.1 μg/cm^2). Furthermore, thrombin enhancement of platelet accumulation was inhibited by antibodies against human platelet adhesive α-granule proteins, in particular vWF and fibrinogen, indicating the role of secreted α-granule proteins in thrombin stimulated platelet deposition.

Activation of the coagulation system in flowing blood interacting with collagen coated glass capillaries has been evaluated by assay of FPA (Ref. 8). In non-anticoagulated whole blood, FPA generation increased with rising shear rate and paralleled platelet deposition, whereas in heparinized blood (10 IU/ml) FPA generation was inhibited and platelet accumulation remained essentially unchanged.

Recently, the perfusion system has been adapted to the study of hollow fibre membranes for hemodialysis. Measurements of protein adsorption from plasma containing traces of radiolabeled proteins indicated adsorption of high surface concentrations (0.5-2.5 $\mu g/cm^2$) of albumin, fibrinogen and immunoglobulin G on both polyacrylonitrile and polysulfone membranes. Preadsorption with human albumin led to low levels of platelet deposition on all types of hollow fibre. Plasma preadsorption also passivated poly-acrylonitrile and cuprophane fibres but appeared to have a platelet activating effect on polysulfone fibres, while fibrinogen preadsorption enhanced platelet deposition moderately on cuprophane and polysulfone and strongly on polyacrylonitrile.

Another line of research in our laboratories concerns the development of techniques for human endothelial cell culture on the inner surface of small calibre (≤ 6 mm diameter) vascular grafts, in order to improve the patency of such grafts in surgical implants (Ref. 9). Biomaterials (expanded polytetrafluoroethylene (ePTFE) or Dacron) were tested in the form of small membranes lining the lower surface of culture dishes. After precoating with human fibronectin, factor VIII concentrate or biological glue containing a mixture of fibrinogen, fibronectin, vWF and factor XIII polymerized with thrombin (Transglutine (TGL, Centre Regional de Transfusion Sanguine de Strasbourg), human saphenous vein endothelial cells adhered to both test materials.

However, cell growth occurred only after pretreatment with biological glue. As compared to the control material, tissue culture polystyrene (TCP) coated with adhesive protein, cell proliferation rates and final cell densities at post confluence were similar on ePTFE/TGL but lower on Dacron/TGL surfaces: confluent density (cells/cm² $\times$ 10^{-3}) 66.4 ± 8.7 (TCP/TGL) 59.2 ± 4.9 (ePTFE/TGL); 23.5 ± 2.6 (Dacron/TGL). Monolayers of cells on the protein treated biomaterials remained intact in vitro for at least 15 days, exhibited typical endothelial contact inhibited cobblestone pattern and showed little or no reactivity in platelet adhesion tests.

Preliminary experiments were also carried out under flow conditions. Cells were seeded onto ePTFE grafts (4 mm diameter) coated with TGL and after an attachment period of 30 minutes the grafts were submitted to a culture medium flow equivalent to arterial shear rates (700 s⁻¹) for two hours. Scanning electron microscopic examination showed 88% cell adhesion and progressive cell spreading under flow without notable cell loss. These results suggest that endothelial cells available from small fragments of human veins could provide an ideal non-thrombogenic lining for small calibre vascular grafts, pretreatment of these vessels with adhesive proteins such as TGL allowing adequate growth conditions and good cohesion between endothelial monolayers and the biomaterial.

References

1. Salzman, E.W. and Merrill, E.W., 'Interaction of blood with artificial surfaces', in: *Hemostasis and Thrombosis. Basic Principles and Clinical Practice*, (2nd ed.) Colman, R.W., Hirsh, J., Marder, V.J. and Salzman, E.W. (Eds). J.B. Lippincott Company, Philadelphia, 1987, 1335-47.
2. Horbett, T.A. and Brash, J.L., 'Proteins at interfaces: current issues and future prospects', in: *Proteins at Interfaces: Physicochemical and Biochemical Studies*, Brash, J.L. and Horbett, T.A. (Eds). American Chemical Society, Washington D.C., 1987, 1-33.
3. Cazenave, J.P., 'Interaction of platelets with surfaces', in: *Blood-surface Interactions. Biological Principles Underlying Haemocompatibility with Artificial Materials*, Cazenave, J.P., Davies, J.A., Kazatchkine, M.D. and van Aken, W.G. (Eds). Elsevier, Amsterdam, 1986, 89-105.
4. Mulvihill, J.N., Huisman, H.G., Cazenave, J.P., van Mourik, J.A. and van Aken, W.G., 'The use of monoclonal antibodies to human platelet membrane glycoprotein IIb-IIIa to quantitate platelet deposition on artificial surfaces', *Thromb. Haemostas.* 1987, 58: 724-31.
5. Mulvihill, J.N., Faradji, A., Oberling, F. and Cazenave, J.P., 'Surface passivation by human albumin of plasmapheresis circuits reduces platelet accumulation and thrombus formation. Experimental and clinical studies', *J. Biomed. Mater. Res.* 1990, 24, 155-63.
6. Cazenave, J.P. and Mulvihill, J.N., 'Capillary perfusion system for quantitative evaluation of protein adsorption and platelet adhesion to artificial surfaces', in: *Proteins at Interfaces: Physicochemical and Biochemical Studies*, Brash, J.L. and Horbett, T.A. (Eds). American Chemical Society, Washington D.C. 1987, 537-50.
7. Mulvihill, J.N., Davies, J.A., Toti, F., Freyssinet, J.-M. and Cazenave, J.P., 'Thrombin stimulated platelet accumulation on protein coated glass capillaries: role of adhesive platelet α-granule proteins', *Thromb. Haemostas* 1989, 62: 989-95.
8. Cazenave, J.P. and Mulvihill, J.N., 'Interactions of blood with surfaces: hemocompatibility and thromboresistance of biomaterials', *Contr. Nephrol.* 1988, 62: 118-27.
9. Mazzucotelli, J.P., Klein-Soyer, C., Beretz, A., Neumann, M.R., Kieney, R. and Cazenave, J.P., 'Improvement of endothelialization of artificial vascular grafts in vitro by precoating with Transglutine , a biological glue', in: *Hybrid Artificial Organs*, Baquey, C. and Dupuy, B. (Eds). Colloque INSERM, 1989, 177: 293-300.
10. Blomback, B. and Blomback, M., *Ark. Kemi.* 1956, 10, 415-443.
11. Kekwick, R.A. et al., *Biochem. J.* 1955, 60, 671-683.
12. Engvall, E. and Ruoslahti, E., *Int. J. Cancer.* 1977, 20, 1-5.
13. Regoeczi, E., *Iodine-Labelled Plasma Proteins.* CRC Press, Florida, 1984.
14. Buchanan, M.R. et al., *Thromb. Res.* 1979, 16, 551-555.
15. Mulvihill, J.N. and Cazenave, J.-P., Proceedings ESAO, Life Support Systems, 1982, 287-290.
16. Hatton, M.W.C. et al., *J. Lab. Clin. Med.* 1980, 96, 861-870.
17. Hemmendinger, S. et al., *Nephron* 1989, 53, 147-151.

7. Protein adsorption at polymer-liquid interfaces using series of polymers with varying hydrophilicity, charge and chain mobility

H.S. VAN DAMME and J. FEIJEN

University of Twente, Department of Chemical Engineering, Enschede, The Netherlands

Introduction

The first event after contact of polymeric surfaces with blood is the adsorption of proteins at the solid-liquid interface. Thereafter processes like the activation of the intrinsic coagulation, adhesion and aggregation of platelets and the activation of the complement system may take place, depending on the composition of the adsorbed protein layer and the conformation of the adsorbed proteins (Fig. 1). The composition of the adsorbed protein layer usually changes as a function of the exposure time.

To obtain more insight in the relation between the character of the polymer surface and it's blood compatibility, protein adsorption has to be studied on series of polymers with well characterized surface structures.

Several types of interactions between the protein and surface such as hydropholic, ionic and donor-acceptor interactions may take place in (Fig. 2). These interactions also determine whether changes in the conformation of the protein molecules will occur upon or after adsorption.

The aim of this study is to get more insight in the influence of polymer surface

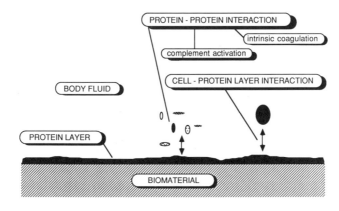

Fig. 1. The role of adsorbed proteins in body fluid biomaterial interaction.

Y.F. Missirlis and W. Lemm (eds), Modern Aspects of Protein Adsorption on Biomaterials, 55-61.
© 1991 *Kluwer Academic Publishers. Printed in the Netherlands.*

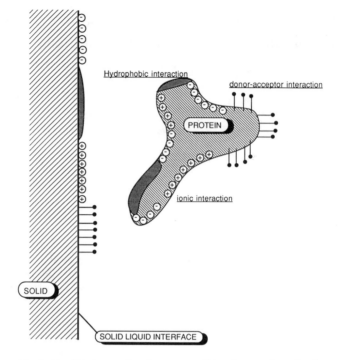

Fig. 2. Possible interactions between a solid and a protein molecule.

characteristics such as hydrophilicity, hydrophobicity, charge and chain mobility of the adsorption of proteins from different protein solutions including plasma. For this purpose well defined polymer surfaces have to be synthesized and characterized, protein adsorption has to be studied and models for the adsorption/desorption process have to be derived (Fig. 3).

Polymer surfaces and protein adsorption experiments

The effect of the side chain length onto protein adsorption was studied using a series of poly(n-alkyl methacrylates) PAMA (Fig. 4). The length of the side chain was varied from n = 1 to n = 18. The PAMA surfaces were characterized by measuring contact angles with the Wilhelmy plate technique. It was found that at room temperature the hydrophobicity of the surface increased from PAMA (n = 1) to PAMA (n = 8) due to the longer hydrophobic alkyl chain. Above n = 8 a decreasing receding contact angle was observed which could be explained by a higher mobility of groups and segments at the polymer surface (Ref. 1). Protein adsorption onto these surfaces was measured from solution with labeled proteins and from plasma with a two step enzyme immuno-assay (EIA).

In another series the effect of surface charge onto protein was studied using

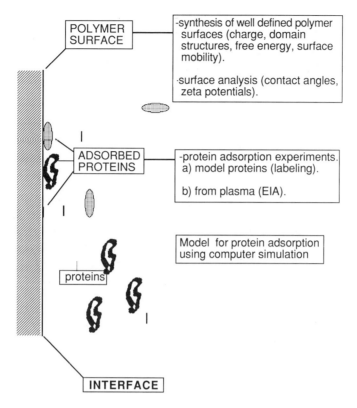

Fig. 3. General setup of the study.

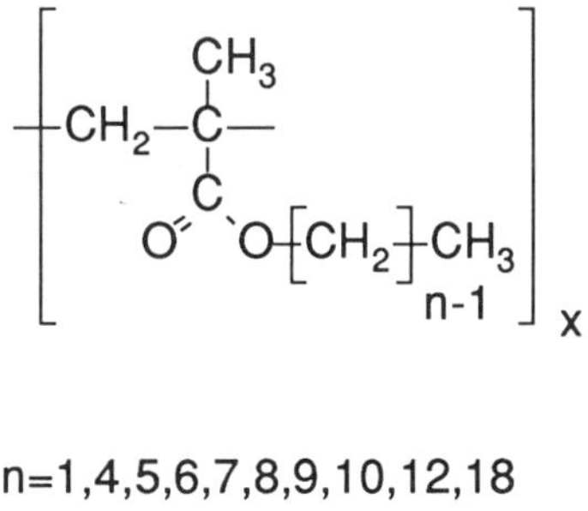

$$\left[\begin{array}{c} CH_3 \\ +CH_2-C- \\ \underset{O}{\overset{C}{\diagup}} \underset{}{\diagdown} O + CH_2 + CH_3 \\ n-1 \end{array} \right]_x$$

n=1,4,5,6,7,8,9,10,12,18

Fig. 4. Repeating unit of poly(n-alkyl methacrylates).

copolymers of methyl methacrylate (MMA) with either methacrylic acid (MAA) or with trimethylaminoethyl methacrylate (TMAEMA) (Fig. 5). Copolymerization with MAA gives more negatively charged surfaces while TMAEMA gives more positively charged surfaces as compared to the pure MMA polymer. In this case protein adsorption was studied from diluted plasma using the EIA technique.

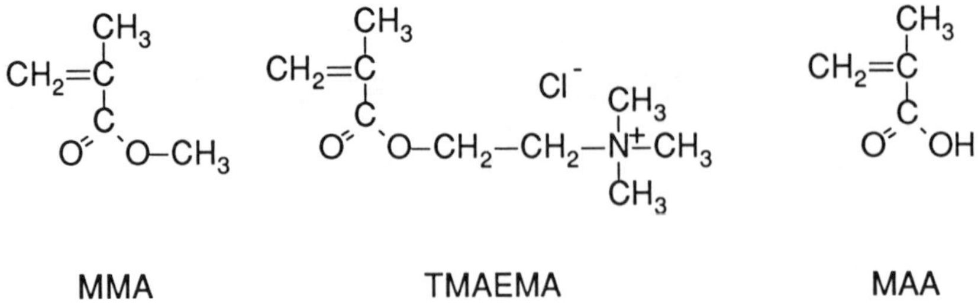

58

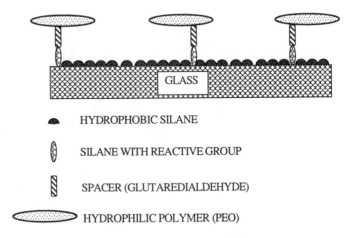

CH₃ image at top — three monomer structures labeled below

MMA   TMAEMA   MAA

Fig. 5. Monomers used in the copolymerization for charged polymers.

A third series of surfaces was prepared to study the effect of the ratio of hydrophilic and hydrophobic groups onto protein adsorption. The aim was to obtain hydrophilic/hydropholic area's with a size comparable to the size of a protein molecule. These structured surfaces were synthesized by grafting of polyethylene oxide (PEO) onto silanized surfaces (Fig. 6). The PEO molecules provide the hydrophilic area's while the hydrophobic area's consist of trimethyl silyl groups. The size and the amount of the hydrophilic groups is controlled by the length of the PEO molecule and the amount of coupling sites for the PEO molecules respectively. The surfaces were analyzed by measuring the amount of PEO molecules on the surface using radiolabeled compounds. The distribution of the groups was checked with auto-radiography. The surfaces were also characterized with contact angle measurements. The competitive adsorption of albumin and fibrinogen onto these surfaces was studied with a dual label technique.

GLASS

- HYDROPHOBIC SILANE

- SILANE WITH REACTIVE GROUP

- SPACER (GLUTAREDIALDEHYDE)

- HYDROPHILIC POLYMER (PEO)

Fig. 6. Structure of the series of surfaces with hydrophilic and hydrophobic domains.

Results

Adsorption isotherms of ^{14}C labeled albumin (HSA) onto PAMA (n = 1, 6, 8 and 18) were not significantly different. In all cases a logarithmic relation between the concentration and the adsorbed amount was observed (Temkin isotherm). Desorption of HSA preadsorbed onto PAMA (n = 1, 8 and 18) after contact with PBS (phosphate buffered saline) for periods up to 48 hours did not take place.

When PAMA surfaces with preadsorbed HSA were exposed to either solutions of non-labeled HSA or fibrinogen (Fg), exchange of preadsorbed HSA with protein in the solutions was observed. The exchange rate of PAMA (n = 1) was lower as compared to that on PAMA(n = 8 and n = 18). Although differences in exchange rates of PAMA (n = 1) and PAMA (n = 8, n = 18) were always observed, a lower HSA concentration in the solution leads to lower exchange rates. Similar differences were found when Fg was used to exchange the preadsorbed HSA.

Adsorption of proteins from diluted plasma onto PAMA (n = 1, 8 and 18) was studied using EIA. Adsorption of HSA, Fg, HDL (high density lipoprotein) and HMWK (high molecular weight kininogen) was studied as a function of the plasma dilution after 1 hour adsorption and as a function of time at two plasma dilutions (2 and 1000).

In the experiments as a function of the plasma dilution higher amounts of adsorbed HSA and HDL were observed on PAMA (n = 8) as compared to PAMA (n = 1 and 18). The amounts of adsorbed Fg and HMWK were the same for all three surfaces. Plateau values were observed for HDL on all surfaces and for HSA on PAMA (n = 8). In all other cases a maximum adsorption was observed at intermediate dilution (1000-10000).

High levels of HSA and HDL adsorbed on PAMA(n = 8) were also observed when the adsorption was carried out as a function of time. The amount of HDL adsorbed increased in time at all surfaces. The amount of adsorbed Fg and HSA did not change significantly as a function of time.

The adsorption of HSA, Fg, IgG (Immunoglobulin G) and HDL at the surfaces of the charged methacrylic copolymers was measured as a function of plasma dilution with the EIA technique.

The results for HSA were not reproducible so no conclusion could be made about the effect of the surface charge on HSA adsorption. The results for Fg on the other hand were very reproducible. No significant difference between the various surfaces could be detected. In all cases the amounts of Fg were maximal at intermediate plasma dilution (100-100000).

For IgG large differences in the amount adsorbed at the different surfaces were found in the high dilution range. However, no simple relation between the adsorbed amount of IgG and the surface charge was found. In the case of HDL, plateau values were found for all copolymers. The plateau value for the copolymer of MMA with MAA was significantly lower than the other plateau values. A straightforward relation was observed between the surface charge and the dilution where the plateau value was reached. The more positive the surface charge the higher the dilution where the plateau value was reached which indicates high affinity of the positive surfaces for HDL.

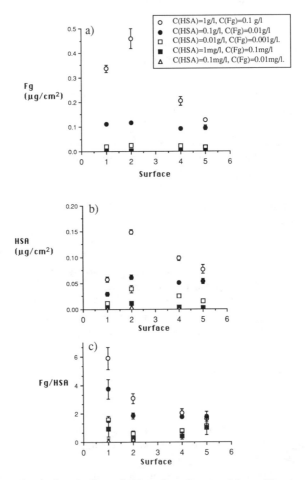

Fig. 7. Adsorption isotherms of competitive adsorption, t = 1 hour. The amount of PEO-molecules on the surface is decreasing with increasing surface number.
a) Fibrinogen. b) Albumin. c) Fg/HSA ratio.

EFFECT OF THE RATIO OF HYDROPHILIC AND HYDROPHOBIC GROUPS

Competitive adsorption of fibrinogen and albumin onto surfaces with different ratio's of hydrophilic and hydrophobic areas was measured using ^{3}H labeled fibrinogen and ^{14}C labeled albumin. Adsorption was measured as a function of the concentration after 1 hour (Fig. 7).

In these experiments the ratio of hydrophilic/hycrophobic areas is decreasing with increasing number. Fig. 7 shows that a maximal amount of adsorbed albumin and fibrinogen at the highest concentration in solution was found onto surface 2 containing an intermediate amount of PEO molecules.

Besides these experiments at a fixed time, three other competitive adsorption experiments were performed using the same surfaces. First, competitive adsorption was studied as a function of time (kinetics). Second, adsorption was measured at a fixed concentration of one of the proteins while the other protein concentration was varied. Finally adsorption of fibrinogen onto surfaces with preadsorbed albumin was studied.

In the kinetic experiments HSA adsorption was maximal on the most hydrophobic surface (surface 5) after adsorption times longer than 2 hours. Both in the experiments as a function of concentration and the exchange experiments the highest adsorption of HSA was observed on surface 2. In all experiments high affinity adsorption of Fg was observed onto surface 2 probably due to strong interaction between this protein and the amine groups terminating the PEO chain

Reference

1. Damme, H.S. van, Hogt, A.H. and Feijen, J., *J. Colloid Interface Sci.* 1986, 114, 167.

8. Cooperative protein adsorption on surfaces with controlled alkyl-residue lattices

H.P. JENNISSEN

Institut für Physiologische Chemie, Universität GHS, Essen, Germany

Introduction

We are all aware of the fact that in mammalian organisms the first reaction which occurs when a foreign surface is introduced into the body is protein adsorption. Besides determining whether a thrombogenic or non-thrombogenic reaction will occur, this protein on the surface is also decisive for another very important reaction, namely, whether bacteria will adsorb to the surface or not, endangering the organism e.g. by a recurrent bacterial dissemination.

It is the aim of our work to gain insight into the fundamental mechanism of protein adsorption, so that this knowledge can be used for the tailoring of biomaterials.

Materials and methods

For the preparation of substituted adsorbents we employed a beaded form of the hydrogel agarose (Sepharose 4B, Pharmacia, Uppsala) as inert, hydrophilic matrix. Agarose consists of a hierarchy of structures: primary (structure: agarobiose units), secondary (structure: single helix), tertiary (structure: double helix), quaternary structure: stacked double helices (Ref. 1, 2). This stacking of double helices can lead to the formation of two-dimensional sheets of agarose thus providing a two-dimensional surface for protein adsorption. The mean pore radius of Sepharose 4B has been estimated to be ca. 80 nm (Ref. 3). This value has recently been confirmed by scanning electron microscopy of Sepharose 4B (Ref. 4). The ratio of pore radius to molecular radius of phosphorylase *b* is ca. 14:1 (Ref. 3), illustrating that restrictions in diffusion and a non-planarity of the surface due to pore curvature can be neglected.

SYNTHESIS OF CONTROLLED SURFACES

Agarose is polygalactose containing primary and secondary alcoholic hydroxyl groups. The primary hydroxyl is the one which is reactive in most substitution

Y.F. Missirlis and W. Lemm (eds), Modern Aspects of Protein Adsorption on Biomaterials, 63-71.
© 1991 *Kluwer Academic Publishers. Printed in the Netherlands.*

reactions. As methods for the coupling of aliphatic amines or mercaptans (C1-C6) to the surface of such gels we have used the CNBR method of Porath et al. (Ref. 5) (carbamate and isourea linkages) leading to alkyl-N_I-agaroses (Ref. 6, 7), the carbonyl diimidazole method of Bethel et al. (Ref. 8) (carbamate linkage) leading to alkyl-N_{II}-agaroses (Ref. 4) and the tresyl chloride method (Ref. 9) (thioether linkage) leading to alkyl-S-agaroses (Ref. 4,10). These latter gels have been shown to be very effective adsorbents of fibrinogen (A. Demiroglou and H.P. Jennissen, *Biol. Chem.* Hoppe-Seyler 371, 778-779 (1990).

The immobilized alkyl residue surface concentration was determined by the addition of tracers of ^{14}C-alkylamines (Ref. 6, 7, 11) or ^{14}C-alkylmercaptan (Ref. 4, 10) to the alkyl containing solution employed for covalent coupling to the agarose matrix. Surfaces with variable surface concentrations of immobilized alkyl residues between 12-350 nmol/m^2 can easily be synthesized in this way. Conditions were chosen where volume changes of the gel either as a function of the degree of substitution or the temperature are very small (Ref. 7, 10). An enhancement of the rigidity of the gel is often obtained by crosslinks introduced into the gel as a result of the activation procedure (also see Ref. 5). The rigidity also makes changes in surface structure as a result of protein binding improbable.

The alkyl-S-derivatives have about a 10-20 fold higher affinity in binding proteins as do their alkyl-N-counterparts of the same chain-length and surface concentration (Ref. 4, 10). Experiments of this type demonstrate that proteins do not only recognize the tip of immobilized residues (e.g. terminal methyl group) but also the atoms at the base of the residue (i.e. nitrogen or sulfur). For this phenomenon the term 'base-atom recognition' is suggested.

Mainly we have employed the protein phosphorylase *b* (molecular mass of 197 kDa) as adsorbate or ligand. It is a dimer comprising a rectangular prism of the size of 11.6 nm x 6.3 nm x 6.3 nm (Ref. 12). This enzyme can be labelled with tritium (^{3}H-Phosphorylase b_r) according to Ref. 13. In the labelling procedure the azomethine bond between Lys 679 and pyridoxal phosphate (prosthetic group of phosphorylase) is reduced either by $NaBT_4$ (labelled enzyme) or $NaBH_4$ (cold-labelled enzyme) thus leading to practically identical structures of labelled and unlabelled enzyme. In comparison to phosphorylase *b* fibrinogen is a fibrillar protein with 1.6 fold molecular weight and a fourfold length of about 45 nm.

MEASUREMENT OF PROTEIN ADSORPTION AND SORPTION KINETICS

As a generally applicable method for the measurement of protein sorption on particulate materials we have introduced the grid-syringe method (Ref. 3, 6, 14). In this method samples are taken with a syringe, which is tipped with a porous grid (pore diameter about 10-20 μm). In the gel-free bulk liquid obtained in this way the free bulk protein concentration and enzyme activity can be easily determined (Ref. 6). With this method kinetics can also be measured with a time resolution of 5-10 seconds (Ref. 14). For adsorption isotherms the amount of adsorbed protein is determined by the depletion method (measurement of a decrease in free concentration of bulk ligand, due to adsorption) (Ref. 6). Conversely in the case of desorption isotherms protein is desorbed by dilution of the enzyme-alkyl agarose complex and determined by the repletion method (measurement of increase in free concentration of bulk ligand released by desorption) (Ref. 3). All binding measurements were performed at high ionic strength (1.1 M ammonium sulfate) to reduce charge effects.

Previously (Ref. 3, 6) it was shown that apparent equilibria of adsorption and desorption are obtained within 60-120 minutes. In our studies it has been stressed that the obtained 'equilibria' do not correspond to true equilibria because of the hysteretic nature of adsorption (Ref. 3, 16) (kinetic hypothesis of protein adsorption).

The evaluation of adsorption data has also been extensively discussed in Ref. 16.

Review and discussion

PROTEIN ADSORPTION ISOTHERMS

In general two categories of protein adsorption isotherms, which encompass cooperative binding, can be distinguished on substituted surfaces (see Scheme 1) (Ref. 16, 17): (a) the lattice-site binding function, and (b) the bulk ligand binding function.

Lattice-site binding function

The lattice site binding function (Ref. 16, 17) describes the binding of surface-immobilized residues to a protein (saturation of the protein surface, not adsorbent surface, with immobilized alkyl residues). The free protein concentration is kept constant (Ref. 17). Several types of isotherms are possible under these conditions.

In the first type, independent binding on single residues leads to hyperbolic isotherms of the Langmuir-type. The reversible adsorption of a rigid protein (absence of significant conformational changes) to single independent sites spaced so far apart that lateral interactions are excluded should yield isotherms of this type (see Ref. 16 for extensive discussion).

In the second type of isotherm in this category cooperative binding to multiple residues leads to non-hyperbolic isotherms of the sigmoid type. These sigmoidal type isotherms are power functions and can be linearized in double logarithmic coordinates either directly or according to eq. 1 in Scheme 1. An

1. LATTICE-SITE BINDING FUNCTION
(Saturation of **protein surface** with immobilized residues)

$$\frac{\theta_S}{1 - \theta_S} = K_S \, (\Gamma_r^s)^{\, n_S} \qquad\qquad (1)$$

Surface adsorption coefficient (n_S):

For $n_S = 1$: <u>Hyperbolic Isotherm</u> (Langmuir Isotherm)
  – 1 lattice-site/protein molecule
  (monovalent adsorption)

For $n_S \geq 1$: <u>Non-hyperbolic Isotherm</u> (sigmoidal type; positive cooperativity
  – multiple lattice-sites/protein molecule
  (oligo-/multivalent adsorption)

2. BULK LIGAND BINDING FUNCTION
(Saturation of **gel surface** with protein)

$$\frac{\theta_B}{1 - \theta_B} = K_B \, [\, P\,]^{\, n_B} \qquad\qquad (2)$$

Bulk adsorption coefficient (n_B):

For $n_B = 1$: <u>Hyperbolic Isotherm</u> (Langmuir Isotherm)

For $n_B \leq 1$: <u>Non-hyperbolic Isotherm</u> (Freundlich Type; negative cooperativity
  – saturation dependent decrease in valence
  – intermoleculare repulsion

For $n_B \geq 1$: <u>Non-hyperbolic Isotherm</u> (sigmoidal type; positive cooperativity
  – intermolecular attraction

Scheme 1. Categories of adsorption isotherms on alkyl-substituted agarose gels. θ_S and θ_B correspond to the respective fractional saturation (Ref. 17) n_S and n_B to the surface and bulk adsorption coefficients and K_S and K_B to the surface and bulk adsorption constants respectively from which the half-saturation constants can be derived. Γ_r^s and [P] correspond to the surface concentration of immobilized residues and to the free protein concentration at equilibrium respectively.

elongation of the alkyl residue (i.e. increase in binding affinity of single residue interaction) leads to a left shift of the sigmoidal curves and to a decrease in sigmoidicity (Fig. 1). This applies to homogenous binding-site lattices (alkyl residues alone) as well as to heterogenous lattices (alkyl-residues plus charges).

In isotherms of the sigmoid type (Fig. 1) adsorption begins at a 'threshold'

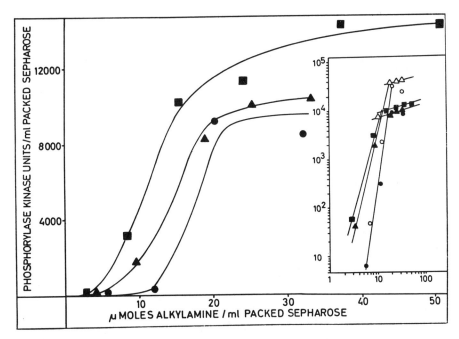

Fig. 1. Adsorption of immomilized alkyl residues to phoshorylase kinase. The enzyme was adsorbed from crude extract (solid symbols) and in purified form (open symbols) to 10 ml packed alkyl agarose in a column. The insert illustrates that the curves are power functions. (○, ●) methyl-N-Sepharose, (△, ▲) ethyl-N-Sepharose, (■) butyl-N-Sepharose. Data from Ref. 18.

(Ref. 18) value i.e. a critical surface concentration of immobilized residues (critical hydrophobicity). The threshold may correspond to a critical adsorption energy postulated for polymer adsorption (Ref. 19). The sigmoidal lattice-site binding curves result from an enhancement of binding affinity of the gel surface for the protein as a result of the increase in surface concentration of immobilized residues (Ref. 3, 6, 16). The surface adsorption coefficient (n_s) allows an estimation of the minimum number of valences or contact sites (i.e. alkyl-residue protein interactions) which are involved in binding (Ref. 3, 16, 17). This interaction coefficient (n_s) is also influenced by lateral interactions and conformational-change dependent interactions with surface sites.

Other forms of non-hyperbolic isotherms in the category of the lattice-site binding function can be expected at very high surface concentrations of immobilized residues (Ref. 16) and with very long immobilized residues (e.g. C8-C12), when a reaction of the immobilized residues with one another e.g. to immobilized micelles is possible. Since residue-area is excluded from interaction with the protein under these conditions the degree of substitution will not reflect the effective surface concentration of immobilized residues.

Bulk ligand binding function

The bulk ligand binding function (Ref. 16, 17) which corresponds to the classical adsorption isotherm monitors the saturation of the adsorbent surface as a function of the free equilibrium protein concentration. The surface concentration of immobilized residues is kept constant.

Hyperbolic Isotherms of the Langmuir Type can be expected under similar conditions as described above for the lattice site binding function (Ref. 16, 17).

Non-hyperbolic protein adsorption isotherms of the *Freundlich type* were first reported in 1976 (Ref. 6). The isotherms were analyzed according to the Scatchard and Hill equations which are the basis of the equations in scheme 1 (see eq. 2 in Scheme 1). As a result of this analysis it was found that the deviation of the Freundlich isotherm from the classical Langmuir type is due to a continuous decrease in binding affinity as a function of fractional gel saturation. This has been termed negative cooperativity (Ref. 6). Why does the affinity decrease as the surface is covered? The number of immobilized alkyl residues interacting with the protein (i.e. valence, contact sites) decreases systematically as the surface becomes saturated with protein. As a result of binding unit overlap (Ref. 3, 6, 14) the number of available free immobilized residues is sequentially reduced (Ref. 6). Additionally as the packing of protein molecules gets higher the optimal geometry allowing certain surface configurations is also reduced (Ref. 6, 14). Depending on the conditions repulsive electrostatic (low salt concentrations) and non-electrostatic repulsive forces (e.g. repulsive van der Waals forces, repulsive hydration forces, steric repulsion, see Ref. 20) may become significant especially when the packing gets high. Repulsive collisional forces might also occur if surface diffusion is significant (Ref. 14). In overlapping collisional areas there may also be a competitive take-over of alkyl residues from one protein molecule to the next (Ref. 6). In addition adsorption is influenced by changes in configuration and conformation (i.e. entropy of binding, Ref. 6). In sum all of these processes lead to a decrease in affinity as expressed in the Freundlich type isotherm (Ref. 6, 14).

Adsorption isotherms of the Freundlich type on various surfaces have also been demonstrated for fibrinogen (Ref. 21-23), albumin and hemoglobin (Ref. 21) adsorption and therefore do not appear to be specific for proteins or surfaces.

Theoretically non-hyperbolic isotherms of the sigmoid type can also occur in this category. These adsorption isotherms have been discussed in Ref. 16 and may occur if intermolecular attraction leads to an aggregation of protein molecules on the gel.

ADSORPTION HYSTERESIS

The inability to detect protein desorption is often interpreted as 'irreversibility' of binding. Since all spontaneous reactions are thermodynamically irreversible the use of the term 'irreversible' for non-detectibility of desorbed protein is very problematical. Non-detectibility only reflects high affinity and non-optimal tools for detection. We have therefore suggested (Ref. 14, 16) that the term 'irreversible' be reserved for thermodynamically irreversible phenomena which display irreversible entropy production ($\Delta_i S$) as has been shown for protein adsorption hysteresis (Ref. 3, 15).

Isotherms in hysteretic systems

In general desorption isotherms have been largely neglected in the past. However, only the measurement of desorption isotherms leads to the detection of adsorption hysteresis.

Protein adsorption (phosphorylase b) on alkyl agaroses is distinctive (Ref. 3) by the fact that the desorption isotherms do not retrace the original adsorption isotherm. The two isotherms lead to adsorption-desorption loops (Ref. 3). We found that the different course taken by the desorption isotherm is due to a drastic increase in the affinity of the protein-agarose interaction in comparison to the affinity underlying the adsorption isotherm (Ref. 3). The affinity increase can be understood as an increase in the number of alkyl-residue protein interactions (i.e. valence) after initial adsorption due to changes in the conformation and/or orientation of the protein on the surface thus exposing new hydrophobic sites for interaction with immobilized residues.

From the ability to distinguish the adsorption and desorption isotherms it was concluded that non-equilibrium long-lived metastable states occur and play an important role in protein adsorption hysteresis (Ref. 3).

From the path-dependent process in a hysteresis cycle it was further concluded that the adsorption process may not be thermodynamically but kinetically determined (Ref. 16) (*kinetic hypothesis of protein adsorption*). Metastability can then be explained as a binding reaction which terminates in a local energy minimum instead of the global minimum.

The Gibbs free energy ($\Delta_i G$) of irreversible adsorption on butyl agarose lies between −2 to −10 kJ/mol (Ref. 15) depending on the surface concentration of immobilized residues. Free energy changes of this magnitude are often associated with conformational changes of proteins (see Ref. 15). It has therefore been concluded that the degree of conformational change of phosphorylase b adsorbed to butyl Sepharose increases as a function of the surface concentration of residues. These and other factors are encompassed in the 'surface dynamic hypothesis of protein adsorption hysteresis' (Ref. 15).

SORPTION KINETICS

Adsorption kinetics

An evaluation of the initial adsorption rates of phosphorylase b to butyl Sepharose led to the detection of saturation type kinetics of adsorption (Ref. 7,

14). From these results it was concluded that a rate-limiting step other than the collision rate is decisive for the adsorption of this enzyme in a multistep manner to the surface. This rate-limiting step may be a conformational or orientational change of the enzyme on the surface essential for the initial 'recognition' step of binding which is then followed by an increase in the number of valences or contact sites (Ref. 14).

Desorption kinetics and cooperative displacement

The decrease in affinity observed in binding experiments as a function of fractional surface saturation (negative cooperativity) is also supported by classical and competitive kinetic experiments (Ref. 14, 24).

In the phosphorylase/butyl agarose system it was shown that the desorption rate (off-rate) and the corresponding off-rate constant increase exponentially as a function of the fractional saturation (Ref. 14, 24). Since the binding and kinetic experiments were performed at high ionic strength electrostatic repulsion is very improbable as a cause of the affinity decrease. The increase in the off-rate constant moreover can be attributed to a saturation dependent reduction in the number of valences of the protein-surface interaction (see discussion on negative cooperativity).

Negative cooperative surface kinetics are particularly conspicuous in so-called competitive displacement experiments, a cooperative displacement of adsorbed molecules by newly adsorbed molecules (Ref. 14). This displacement effect is not due to a simple exchange reaction but to the lateral disturbance of the bound state of the molecule adsorbed first by the newly adsorbed second molecule. This disturbance is accompanied by a 4-5 fold increase in the off-rate constant of the first molecule through the second molecule (intermolecular collision, repulsive forces). The model has been extensively described in ref. (Ref. 14, 15).

Conclusion

The concept of cooperativity in protein adsorption allows a plausible explanation for such complex phenomena as sigmoidal isotherms, Freundlich isotherms and the corresponding surface kinetics. It also gives a first understanding of protein adsorption hysteresis. Analysis of these phenomena was to a large extent only possible through a controlled synthesis of alkyl surface lattices.

Acknowledgements

This work was supported by a grant of the Bundesministerium für Forschung und Technologie and the Fonds der Chemie which is gratefully acknowledged.

References

1. D.A. Rees, *Biochem. J.* 126, 257-273 (1972).
2. S. Arnott, A. Fulmer, W.E. Scott, I.C.M. Dea, R. Moorhouse and D.A. Rees, *J. Mol. Biol.* 90, 269-284 (1974).
3. H.P. Jennissen and G. Botzet, *Int. J. Biol. Macromol.* 1, 171-179 (1979).
4. A. Demiroglou, W.J. Kerfin and H.P. Jennissen UCLA Symposia on Molecularand Cellular Biology New Series, Vol. 80, T.W. Hutchens, (Ed.) *Protein Recognition of Immobilized Ligands*, pp. 71-82. Alan R. Liss Inc., New York, 1989.
5. J. Porath, R. Axn, and S. Ernback, *Nature* 215, 1491-1492 (1967).
6. H.P. Jennissen, *Biochemistry* 15, 5683-5692 (1976).
7. H.P. Jennissen, *J. Solid-Phase Biochem.* 4, 151-165 (1979).
8. G.S. Bethell, J.S. Ayers, W.S. Hancock and M.T.W. Hearn, *J. Biol. Chem.* 254,2572-2574 (1979).
9. K. Mosbach and K. Nilsson, *Biochem. Biophys. Res. Commun.* 102,449-457 (1981).
10. A. Demiroglou and H.P. Jennissen, *J. Chromatogr.* 521, 1-17 (1990).
11. H.P. Jennissen, A. Demiroglou and E. Logemann, in: Gribnau, T.C.J., Visser, J. and Nivard, R.J.F. (Eds.), *Affinity Chromatographyand Related Techniques-Theoretical Aspects/ Industrial and Biomedical Applications*, Analytical Chemistry Symposia Series, Vol. 9, pp 39-49. Elsevier, Amsterdam, Oxford, New York, 1982.
12. N.L. Johnson, N.B. Madsen, J. Mosely and K.S. Wilson, *J. Mol. Biol.* 90,703-717 (1974).
13. P.H. Strausbauch, A.B. Kent, J.L. Hedrick and E.H. Fischer, *Methods Enzymol.* 11, 671-675 (1967).
14. H.P. Jennissen, *J. Colloid Interface Sci.* 111, 570-586 (1986).
15. H.P. Jennissen, in: *Handbook Surface and Interfacial Aspects of Biomedical Polymers*, Vol. 2, 'Protein Adsorption', J.D. Andrade (Ed.), pp. 295-320, Plenum Press, New York, 1985.
16. H.P. Jennissen, *Makromol. Chem., Macromol. Symp.* 17, 111-134 (1988).
17. H.P. Jennissen, *J. Chromatogr.* 215, 73-85 (1981).
18. H.P. Jennissen and L.M.G. Heilmeyer, Jr., *Biochemistry* 14, 754-760 (1975).
19. M.A. Cohen Stuart, J.M.H.M. Scheutjens and G.J. Fleer, ACS Symposium Series, No. 240, Goddard, E.D. and Vincent, B. (Eds.) *Polymer Adsorption and Dispersion Stability*, American Chemical Society, pp. 53-65, 1984.
20. J.N. Israelachvili, *Intermolecular Surface Forces*, Academic Press, London, New York, Tokyo, 1985.
21. T.A. Horbett, P.K. Weathersby and A.S. Hoffmann, *J. Bioeng.* 1, 61 (1977) .
22. P.K. Weathersby, T.A. Horbett and A.S. Hoffman, *J. Bioeng.* 1, 395-410 (1977).
23. A. Schmitt, R. Varoqui, S. Uniyal, J.L. Brash and C. Pusineri, *J. Colloid Interface Sci.* 92, 25-34 (1983).
24. H.P. Jennissen, *Adv. Enzyme Regul.* 19, 377-406 (1981).
25. Chafouleas, I.G., Dedman, J.R., Munjaal, R.P. and Means, A.R., *J. Biol. Chem.* 254, 10262-10267 (1979).
26. Cohen, P., *Eur. J. Biochem.* 34, 1-14 (1973).
27. Graves, D.J., Hayakawa, T., Horowitz, A., Beckmann, E. and Krebs, E.G., *Biochemistry* 12, 580-585 (1973).
28. Weigel, P.H. Schnaar, R.C., Kuhlenschmidt, M.S., Schnell, E., Lee, R.T., Lee, Y.C. and Rosemann, S., *J. Biol. Chem.* 254, 10830-10838 (1979).

9. Protein adsorption tests for polymer surfaces

W. LEMM

Rudolf Virchow Clinic, Department of Experimental Surgery, Berlin, Germany

Introduction

Within a national research project of the German Ministry of Research and Technology (BMFT) in the late seventies in vitro screening methods for the evaluation of the hemocompatibility of synthetic material surfaces have to be developed. Based on the hypothesis of D.J. Lyman, that the preference of a blood contacting surface for albumin is identical with good thromboresistant properties, an in vitro test procedure was introduced to quantify the affinity of several biomaterials to the proteins albumin, fibrinogen, and gamma-globulin.

Methods

These three bovine-proteins were commercial products and not particularly purified. They were dissolved according to their physiological concentrations:
– albumin: 4.5 g/100 ml;
– fibrinogen: 0.3 g/100 ml;
– gamma-globulin: 1.2 g/100 ml;
in buffer solutions (pH = 7.2). All protein solutions were protected with sodium-azide (NaN_3) to prevent bacteriological decompositions. Sodium-citrate was added to the fibrinogen solution to avoid any kind of fibrin polymerisation.

The adsorption kinetics were recorded
– in single protein solutions;
– in competition to each other;
– in bovine plasma.
The radiolabelling techniques with iodine-isotopes was used to identify and to quantify the adsorbed proteins.

Circular samples of each test material, 10 mm in diameter and 0.2 mm thick, were exposed to the slowly stirred protein solutions, removed after specified intervals up to 24 hours, extensively washed in isotonic NaCl-solution and finally placed into a three-channel gamma-counter with linked data processing for specific radioactivity assay.

Y.F. Missirlis and W. Lemm (eds), Modern Aspects of Protein Adsorption on Biomaterials, 73-79.
© 1991 *Kluwer Academic Publishers. Printed in the Netherlands.*

Results

Single protein adsorption

The time depending single protein adsorption of fibrinogen, gamma-globulin, and albumin during a period of 24 hours are shown in Fig. 1-3.
- Polyurethanes have a higher affinity to proteins than silicone or synthetic rubber;
- The adsorption of fibrinogen and gamma-globulin is faster than of albumin;
- Surfaces are saturated with fibrinogen and gamma-globulin after 1-2 hours; but saturation with albumin is completed after 24 hours;
- Prealbuminized surfaces seem to be equalized and reduce the ability to adsorb fibrinogen and gamma-globulin (Fig. 4).

Competitive protein adsorption

A more closer simulation of the in vitro situation are the competitive protein

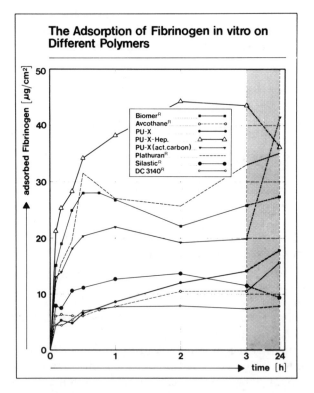

Fig. 1. The adsorption of fibrinogen on different polymers in vitro.

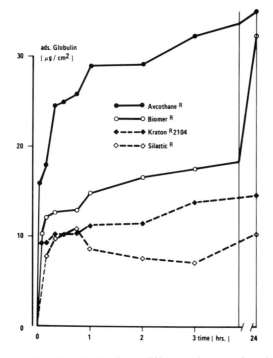

Fig. 2. The adsorption of globulin on different polymer surfaces in vitro.

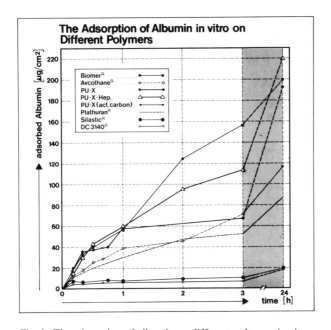

Fig. 3. The adsorption of albumin on different polymers in vitro.

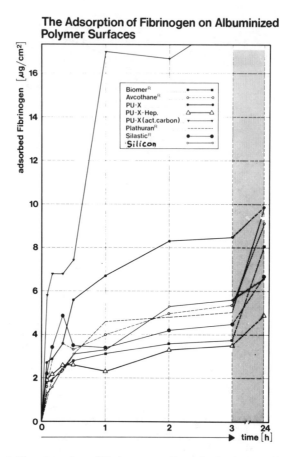

Fig. 4. The adsorption of fibrinogen on albuminized polymer surfaces.

adsorption studies. The surface concentration of fibrinogen is rather unstable during the first 60 minutes; albumin seems to replace recently adsorbed fibrinogen or gamma-globulin (Fig. 5).

Characteristic protein adsorption number (PAN)

The surface affinity to one of the proteins can be expressed quantitatively by characteristic protein adsorption numbers. They are defined as the ratio between the concentration of albumin to fibrinogen, or to gamma-globulin on protein-saturated surfaces (Table 1).

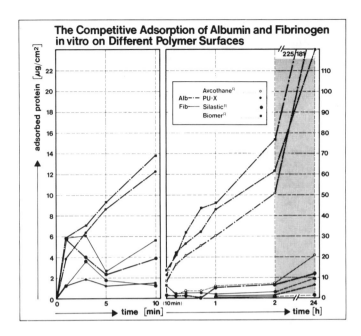

Fig. 5. The competitive adsorption of albumin and fibrinogen on polymer surfaces in vitro.

Table 1. Characteristic Protein Adsorption Numbers of some Biomaterials.

Material	Alb./Fib.	Alb./Glob.
Silastic	2.11	1.90
Silicon 3140 RTV (Coating)	2.26	1.05
Avcothane 51	4.51	3.29
Biomer	7.26	6.14
Pellethane 2363-80A	9.82	5.60
Kraton 2104 (synthet. rubber)	2.13	2.88
Blood	15.00	3.75

Competitive protein adsorption out of plasma

For a further approach to the in vivo conditions adsorption kinetic studies were extended to tests in bovine plasma. The amount of the relevant proteins was controlled by electro-phoreses:
- albumin: 2.29 g/100 ml;
- fibrinogen: 0.477 g/100 ml;
- gamma-globulin: 2.35 g/100 ml.

Basically no essential deviations could be observed comparing the competitive protein adsorption tests in synthetic protein solutions with those in plasma. But the surface concentration of adsorbed albumin is lower in plasma than in synthetic protein solutions. The high concentration of gamma-globulin seems to inhibit the albumin adsorption.

Fibrinogen adsorption in vitro with native blood

To study the adsorption of fibrinogen in blood, the well-known blood-chamber-test invented by Nose was slightly modified. Between two circular sheets of test material 2 ml of native blood of a healthy calf was filled and gently moved. The blood was not anticoagulated! After different time intervals the chambers were opened, the thrombus removed, washed and dried.

The adsorbed quantity of fibrinogen was measured in the same way as described before.

The relation between the surface affinity to fibrinogen and the thrombus formation (weight of dried thrombus) is compared in Table 2.

Table 2. Relation between Fibriogen Adsorption and Thrombus Formation on some Biomaterials.

Silastic	adsorbed fib. after 2 hours [μg/cm^2]	weight of dried thrombus bus after 2 hours [g]
Silastic	9.00	0.62
Pellethane 2363-80A	0.20	0.20
Biomer	0.28	0.33
Plathuran (PEst.-U)	0.16	0.12
Kraton 2104 (rubber)	10.40	0.73

Conclusions

Each of the three test variations helps to explain and to understand the complex events of surface induced thrombus formation.

The simple in vitro studies with synthetic protein solutions yield the results of the high preference of polyurethanes to albumin although the physiological relation between albumin and fibrinogen is not reached. Prealbuminization protects surfaces from fibrinogen adsorption.

The surface affinity to albumin in relation to fibrinogen or gamma-globulin is quantitatively expressed by the characteristic protein adsorption numbers.

The stepwise approach to in vivo conditions contribute to accomplish these basic informations. The presence of gamma-globulin lowers essentially the adsorbed quantities of albumin.

The combination of the protein adsorption test assembly with the blood chamber test from Nose yields a surprising result: the correlation between the adsorbed quantities of fibrinogen and the efficiency of thrombus formation.

The contribution of surface defects to the thrombogenicity of a material cannot be identified by these tests.

References

1. Lyman, D.J., Knutson, K., McNeill, B. and Shibatani, K., *Trans. Amer. Soc. Artif. Organs* 21, 49 (1975).
2. Unger, V. and Lemm, W., *Proceedings of the Europ. Soc. Artif. Organs* 3, 48 (1976).
3. Lemm, W. and Unger, V., *Adsorption of Blood Proteins on Different Polymer Surfaces in vitro*, in: Winter, Leray, de Groot (Eds.), *Evaluation of Biomaterials*, 505-512. John Wiley & Sons Ltd. (1980).
4. Lemm, W., Unger, V., Kaiser, M., Große-Siestrup, Ch., Bucherl, E.S., 'A Comparing Evaluation of three Blood Compatibility Tests for Biomaterials', *Proceedings of the Europ. Soc. Artif. Organs* 5, 29-33 (1978).

10. Advantages and problems using FT-IR spectroscopy to study blood-surface interactions by monitoring the protein adsorption process

A. MAGNANI, M.C. RONCOLINI and R. BARBUCCI

CRISMA, Nuovo Policlinico, Le Scotte, and Dipartimento di Chimica, Università di Siena, 53100 Siena, Italy

The early molecular interactions, that occur when blood contacts a surface, play an important, but incompletely understood, role in the course of thrombogenesis on surface. The objective of this research has always been the monitoring of protein adsorption from blood onto surfaces of biomedical interest, with the intent of correlating the molecular events occurring with the surface chemistry and hemocompatibility. This is perhaps a most ambitious objective, due to the complexity of the competitive adsorption of proteins that occurs when a surface is exposed to blood.

   FT-IR methods were developed to study protein adsorption because it has been felt that this technology has the highest potential for meaningful study of blood-surface interface.

Advantages of FT-IR spectroscopy in the study of blood-surface interaction by monitoring the protein adsorption process

FT-IR instruments employ a Michelson interferometer to produce an interferogram of transmitted or reflected light, whose Fourier transform is the infrared spectrum of the sample. Upon the increase in processing speed of minicomputers, software for fast Fourier transform and the use of improved mercury-cadmium-telluride detectors, interferometers and sample cells, the commercially available FT-IR instruments are now capable of very rapid spectral collection rates, with high signal-to-noise ratios and with simple sample preparation requirements (Ref. 1). Their sensitivity permits the analysis of biological macromolecules in aqueous media, previously hindered by the interference of the strong absorption of the OH-bending vibration of water (centered at 1640 cm^{-1}) that overlaps the region of the strongest absorption bands of proteins (the Amide I band at 1630-1660 cm^{-1}).

   By using these instruments we have the possibility to digitally subtract spectra, obtaining the separation of solvent, solutes and polymer contributions in the combined IR spectrum (Fig. 1; Ref. 2). Other numerical manipulations, such as spectral deconvolution or differentiation, can be used to reveal the

Y.F. Missirlis and W. Lemm (eds), Modern Aspects of Protein Adsorption on Biomaterials, 81-86.
© 1991 *Kluwer Academic Publishers. Printed in the Netherlands.*

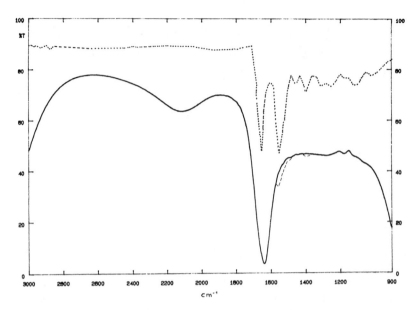

Fig. 1. FT-IR spectra of: –) water, ---) solution, ..) difference = solution – water).

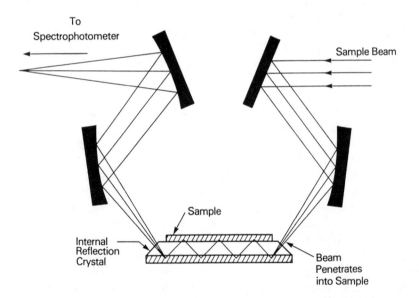

Fig. 2. Typical geometry for Attenuated total reflection spectroscopy.

subtleties of the protein IR features.

When the FT-IR spectroscopy is combined with the Attenuated Total Reflection (ATR) technique (Fig. 2; Ref. 3) a very powerful system to investigate protein adsorption phenomena in static or dynamic conditions is

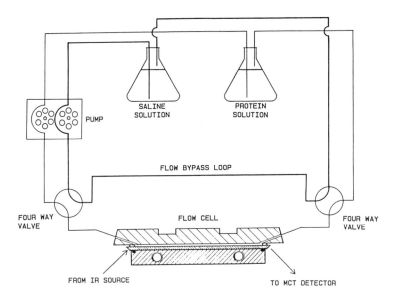

Fig. 3. ATR flow cell and pumpling setup for protein adsorption experiments.

produced. Particular ATR cells are now available to study processes which occur at the 'aqueous medium-material surface' interface (Fig. 3).

Thus, proteins can be characterized on the basis of their intrinsic IR spectra, without the need for extrinsic labels such as fluorescent or radioactive tags.

Problems using FT-IR/ATR to investigate the protein adsorption process

Most of the information about protein adsorption process, obtained by using this technique are qualitative or semiquantitative, owing to the difficulties involved in the analysis of complex IR spectral features for which little or no previous interpretative background data exists.

The technical problems that make quantitative analysis difficult include:

1. the objective subtraction of the water absorption from the protein solutions spectra;
2. the complexity of the FT-IR spectra collected in the Attenuated Total Reflectance mode which can include contributions from: (i) the adsorbed protein layer, and (ii) the soluble proteins in the liquid layer adjacent to the ATR surface but within the 'depth of penetration' of the evanescent field of the IR beam;
3. the similarity of protein spectral recognition based on weaker secondary bands or slight variations in band intensities or frequencies, rather than on strong bands that are unique for each species;
4. the sensitivity of the IR signals of proteins to microenvironmental factors such as pH, ionic strength and the presence of other solutes or surfaces;

5. the inherent complexity of plasma.

The actual research has been directed to assessing the magnitude of these problems and solving them.

Anyway, two important types of information can be derived from FT-IR investigations:

- the rates and amount of adsorption of specific classes of proteins (quantitative aspect) (Ref. 4);
- the conformational changes in the adsorbing species that might indicate a transition of the protein from 'normal' to 'foreign' in terms of recognition by the cascade of thrombosis event (structural aspect) (Ref. 5).

Both these aspects of protein adsorption are important in surface-induced thrombogenesis.

Our effort is so devoted to the analysis of complex protein mixtures in order to accomplish the objective of monitoring the protein adsorption process from whole blood onto polymer surfaces.

It is known that some proteins carry out their functions when dissolved, while others when adsorbed. Because of this predilection of proteins for adsorption, it is important to determine if the assignments for proteins in solution also hold for adsorbed proteins. In other words, is the structure of dissolved and adsorbed protein the same?

Thus, in order to obtain the spectra-structure correlations we have to know the spectrum of the protein in aqueous solution, as well. The frequencies of the infrared peptide backbone vibrations (the so-called Amide vibrations) of proteins can be used to differentiate the conformations in proteins in aqueous solutions.

Table 1 summarizes the data relative to spectra-structure correlation for general proteins.

The Amide I and Amide III are particularly useful to distinguish conformations, but frequencies of one band alone (either Amide I or Amide III) do not yield clear information on conformations.

Table 1. Secondary structure assignments (cm^{-1}).

Protein conformation	Amide I	Amide II	Amide III
α	1656 ± 2(s)	1548 ± 2(s)	1298 ± 4(m)
			1245 ± 3(w)
β	1637 ± 2(s)	1546 ± 8(s)	1235 ± 1(m)
β + Low α	1636 ± 2(s)	1548 ± 4(s)	1254 ± 1(m)
			1236 ± 1(m)
α + Low β	1652(s)	1547 ± 3(s)	1315(m)
			1266 ± 8(w)
			1238(m)
α + β	1644 ± 2(s)	1549 ± 7(s)	1315(w)
			1241 ± 4(m)

s = strong, m = medium, w = weak

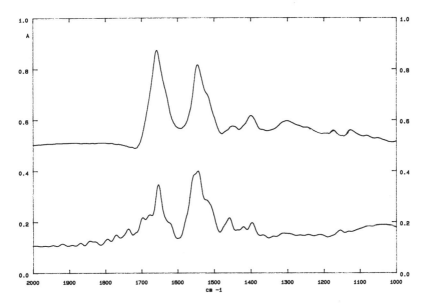

Fig. 4. FT-IR difference spectra of: a) albumin in solution, b) albumin adsorbed onto Ge surface.

Because of the complexity of blood, at the beginning of this study it is necessary to analyse the adsorption process from a single protein solution.

With regard to this, we have been studying the spectra-structure correlation by comparing spectra of adsorbed and dissolved single protein. Fig. 4 shows difference spectra of albumin dissolved in saline (top) and adsorbed on Ge ATR crystal (bottom).

It may be seen that the spectra of the albumin dissolved or adsorbed on Ge appear very similar. Thus, the conformation of albumin adsorbed on germanium surface appears to be identical to that of the protein in solution. Spectra of the adsorbed albumin after being on the surface for over eight hours did not show any changes in the conformational structure.

So, these preliminary results would indicate that adsorption of *Ge surface* produces a protein structure similar to that in solution. However, the initial adsorbed structure may not be the most stable structure and so, rearrangement to a more stable one may occur for some proteins. It is also known (Ref. 6) that the structure of adsorbed albumin, as well as other proteins, changes with the nature of the adsorption surface.

So, the next step in our research program will be the investigation of single protein adsorption onto surfaces of biomedical interest. Then, the adsorption of protein mixtures and proteins from whole blood will be analysed in static and dynamic conditions.

But all these results have to will be compared to others obtained by different techniques, because we believe that any result does not give assurance if it is collected by using only a single technique.

86

References

1. P.R. Griffiths and J.A. deHaseth, *Fourier Transform Infrared Spectroscopy*. J.Wiley & Sons, New York, 1986.
2. J.R. Powell, F.M. Wasacz and R.J. Jakobsen, *Appl. Spectr.* 40, 339, 1986.
3. N.J. Harrick, *Internal Reflection Spectroscopy*. Interscience, New York, 1987.
4. K.K. Chittur, D.J. Fink, R.I. Leininger and T.B. Huston, *J.Colloid Interface Sci.* 111, 419, 1986.
5. S. Winters, R.M. Gendreau, R.I. Leininger and R.J. Jakobsen, *Appl. Spectr.* 36, 404, 1982.
6. R.M. Gendreau and R.J. Jakobsen, *J. Biomed.Mater. Res.* 13, 893, 1979.
7. R.M. Gendreau et al., *Applied Spectroscopy* 35, 353-357, 1981.
8. R.M. Gendreau et al., *Applied Spectroscopy* 36, 47-49, 1982.
9. D.J. Fink et al., *Analytical Biochemistry* 165, 147-154, 1987.

11. Biocompatibility research at Humboldt University

H. WOLF

Abt. exp. Stomatologie und Biomaterialforschung, Humboldt Universität, Berlin, Germany

In our unit we are doing two types of research: Basic research concerning the mechanism of interaction of blood cells with foreign surfaces to understand it in terms of biophysics and also on the cellular or molecular level, concerning the interaction mechanism.

The second topic of our work concerns the development of standardised in vitro methods for evaluation of hemocompatibility and also for in vitro evaluation of bioceramics in soft tissues.

But last, not least, we are included in the development of polymers which are especially used in the artificial heart and in dialysers, so it means, we are looking at special problems of hemocompatibility.

In the field of basic research and especially in protein adsorption we have done some work concerning the question how the surface properties of biomaterials are affected by adsorption of proteins. We have used in our studies techniques coming from colloid chemistry and we have used in principle two methods:

The first method mentioned already by Dr. Van Damme was to make measurements in a flow chamber and to measure the streaming potential or streaming current and so we could characterise surface properties of polymers before and after protein adsorption.

A second technique, which we've used and I want to speak about is the so-called particle electrophoresis. We have used particles or cells in an electric field and as you know the electrophoretic mobility EPM is a function of the surface potential or the so-called zeta-potential, what you can calculate from the Henry Smoluchowsky equation. (Fig. 1 shows as an example the change of EPM of polystyrene latices as function of protein concentration after 15 min adsorption time).

If you use different particles, that means glass, polystyrene or any polymer which is coated on glass you can measure the function of the electrophoretic mobility of the particle as a function of the protein which is added to the electrolyte solution. So you get a change of the surface charge of the particle. For each particular protein you get a special value where the change from the uncoated particle has changed to the fully coated particle. This change of

Y.F. Missirlis and W. Lemm (eds), Modern Aspects of Protein Adsorption on Biomaterials, 87-91.
© 1991 *Kluwer Academic Publishers. Printed in the Netherlands.*

88

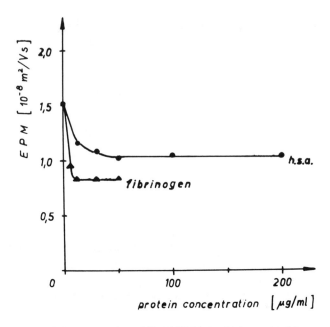

Fig. 1. The change of electrophoretic mobility (EPM) of polystyrene particles as a funtion of protein concentration after 15 min adsorption time at pH 7.4 in phosphate buffered saline solution (PBS) and 25°C.
(h.s.a = human serum albumin)

EPM should not be equal necessarily with the degree of coverage but it is equal to the change of electrical properties.

This method was used in our laboratory especially to investigate competitive protein adsorption. Because we can make precoating of each particle with any protein, we can look how another protein which is in competition with the precoated protein changes the EPM. I will summarise these results here.

If you use for instance fibrinogen then you can change the EPM of this particle as a function of the preadsorption time. The most interesting fact that we did find was the following: if the precoating time of HSA was only maybe some minutes or some hours, then it was possible to replace the HSA. But the longer the precoating time of HSA was, fibrinogen was not able to replace HSA from its binding places. This time was found to be 0.5 h as shown in Fig. 2.

Then we have started experiments for different proteins and we could find that for each type of protein another time was necessary for the precoating of a distinct type of polymer for the replacement of HSA. So you can characterise all the different types of proteins; we've used different IgGs, Gamma-globulin, fibronectin and a lot of other proteins and we've also used different types of polymers. Thus we could find which polymer should be suited maybe for the best so-called irreversible adsorption of HSA. Examples are given in Fig. 2.

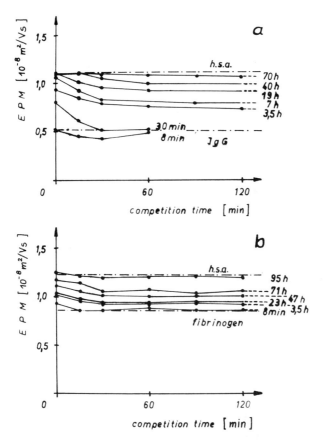

Fig. 2. The competition of IgG (a) or fibrinogen (b) with human serum albumin (h.s.a)
precoating with different precoating times at polystyrene latices.

– . – . – . = dotted lines: EPM of polystyrene particles coated with h.s.a, IgG or fibrinogen
_____ = concentration of IgG or fibrinogen: 25 μg/ml.

Q.(Hoffman): What surfaces are you talking about?

Answer: We've used at first glass, then siliconized glass particles, then we've used polystyrene, latices, or polymer coated glass. It is clear that such a coating procedure, of course, gives you different surface properties.

The interesting point is that you can differentiate each polymer by the electric charge and then you could follow protein adsorption/desorption kinetics for each different polymer. So, I think we have presented results, by which one could find out very easily a material which adsorbs the protein, say irreversibly.

And now: How to explain these results? We have made kinetic models to understand our experimental results and I will not frustrate you with all the equations that are in a paper of my coworker Dr. Mientus (Ref. 2).

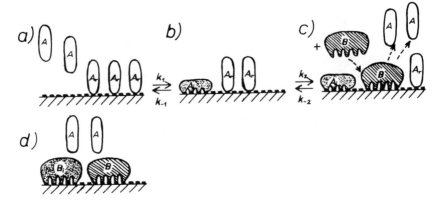

Fig. 3. Kinetic model of competitive protein adsorption onto solid surfaces.
a) Protein molecules of type A are reversibly adsorbed at one binding site to A_r.
b) Reversible adsorbed molecules (A_r) are adsorbed irreversibly with more binding sites to A_i.
c) Protein molecules of type B compete with reversibly adsorbed molecules of type A_r occupying directly more than one binding site.
d) Adsorbed molecules of type B with more binding sites cannot be replaced by molecules of type A.

We have thought about the possible underlying adsorption mechanism and to put this in a kinetic model, which I want to explain schematically in Fig. 3.

We assume that there are binding sites at the solid surface and the protein molecules can interact with one or more binding sites with different adsorption/desorption rates.

Due to our results human serum albumin (HSA) seems to be bound reversibly (Fig. 3a) and to get in an irreversible binding state as a function of interaction time with the solid surface (Fig. 3b) occupying more than one binding site.

A competing molecule B (like fibrinogen) is able to occupy directly more than one binding site and to replace HSA from the surface in case of reversible binding state (Fig. 3c).

On the other side, we have done the opposite experiments, which means, we have precoated the polystyrene particles with fibrinogen, IgG and other molecules. In that case HSA was not able to replace one of those molecule types (Fig. 3d). As Fig. 2 shows the kinetics of HSA replacement depends on the chemical nature of competing molecules, but after a long preadsorption time of HSA, also this molecule seems to be bound irreversibly, which is not believed by John Brash.

The difference between the model which was presented by Sevastianov (Ref. 3) and our model consists in the fact that we have not any diffusion barrier in that model. We are speaking only directly about these binding sites and the advantage of our method is that we have not to label the proteins. We can work without labelling of proteins and of course we perform also experiments with precoating of one protein and using plasma for competition.

References

1. E. Knippel and H. Wolf, 'Anwendung der Festpatikel-Elektrophorese als neuartige, einfache reproduzierbare Methode der Mukoviszidose Diagnose', *Z. Klin. Med.* 42, 1621-1624, 1987.
2. W. Mientus and H. Wolf, *Studies on the kinetics of protein adsorption onto solid surfaces by means of particle electrophoresis*, Proceedings of Symposium 'Electrokinetische Erscheinungen '89', (Editor: Academy of Science of GDR), Dresden 1989, pp. 357-367.
3. V.I. Sevastianov, Z.M. Belomestina and N.K. Zimin, *Artif. Organs* 7, 126-130, 1983.

PART TWO

Discussion chapters

12. Protein characterization

Brash: I think one of the things that we might want to address is the question of protein purity and standards. If two labs work on fibrinogen we should be able to say either it's the same, identically, or we should know what the differences are.

When we work with proteins for adsorption studies, we should be absolutely sure that the proteins are as well characterised as they possibly can be. I think this is possibly a 'motherhood' statement but it's not always adhered to, and maybe something we could address would be minimum criteria of purity in a given protein that we decide we want to work with. Should we be doing only gels or some kind of immunochemical purity measure or what should it be?

Hoffman: There is also the possibility of exchanging and agreeing on protocols for purification of proteins. We have standard polymeric biomaterials that are shared by a number of people around the world. The IUPAC has set up a subcommittee, and I don't know to what extent this is even feasible in the area of protein adsorption.

But if people could agree on specific protocols, this would potentially go a long way towards facilitating and making more meaningful comparisons from one group to another around the world.

Beugeling: I think it depends on the purpose for which you want to use the protein. I will give you an example. In the case of protein precoated surfaces, which were used for instance by Poot (Ch. 4, Ref. 10, 13) in the Cazenave system, he used fibrinogen as a material for precoating. Now, you know that preadsorbed vWF and fibronectin are very important, so for that purpose you must be sure that you have removed these proteins. If there is a little bit of albumin together with fibrinogen it does not matter.

So, I think the purification method depends on the purpose for which you want to use the protein. In case of albumin it was not necessary to purify the protein, because if you take albumin in solution from the blood bank it contains, let's say less than 5% α- and β-globulins and if you precoat the surface with this albumin you don't see any platelet adhesion at all. So, why should you

Y.F. Missirlis and W. Lemm (eds), Modern Aspects of Protein Adsorption on Biomaterials, 95-111.
© 1991 *Kluwer Academic Publishers. Printed in the Netherlands.*

purify your albumin? Do you think that you'll find more platelets or less platelets? I don't think so.

Brash: I think that's well taken but nevertheless regardless of the purpose for which you want to use the protein, my point really is that you should provide some information regarding the status of the protein whether you use it to do adsorption studies or to precoat surfaces or as an antibody in an assay. You should at least provide some measure of the purity of the protein.

And my point really was: can this group make, shall we say, a 'wish list' or even a realistic list of criteria we should all be able to apply. They shouldn't involve esoteric techniques that only a few labs in the world are able to do, but fairly standard, simple techniques that are available in most laboratories.

Beugeling: I fully agree with Prof. Brash that you have to give information about the way you purified your protein and how you characterised it. Poot (Ch. 4, Ref. 10, 13), to mention him again, purified human fibrinogen by chromatography on aminohexyl-Sepharose 4B and characterized it with SDS electrophoresis to see whether he had removed fibronectin and vWF. So, I think SDS-electrophoresis is one of the techniques you should use. That's a rather simple technique.

Cazenave: I think I wouldn't really agree with you and maybe before embarking on looking at what criteria we should all use, we should realise first of all that these criteria have to be applied if we are using proteins which we purify in our labs and that we've also to make the same checks on the proteins which are bought from commercial sources.

When you are looking for example at fibrinogen from Sigma or Kabi, it contains a lot of impurities and these may be very important, depending on the experiment you are doing. And I add and agree to what Tom Beugeling was saying, that if you are looking at platelet adhesion, well, you have to be sure that you have one adhesive protein and not other contaminants. You've to be also sure that your proteins are functional .

We know, for example, that if you add fibrinogen, which is a cofactor of platelet aggregation, if you have traces of thrombin you'll potentiate a reaction which is completely independent of your fibrinogen. Likewise, for most of the proteins of the coagulation or fibrinolytic systems, one has to be sure that you don't have proteolytic fragments. And in addition there is heterogeneity in proteins and there are isoforms or isotypes. So, it might be rather complex but it might be good that we approach this in a discussion.

Baquey: I would say that besides discussions,with protocols used for checking the purity of the proteins which are used, it's very important to check again the purity of the protein after labelling, and to check, as this was mentioned this morning that the labelled protein has the same properties as the native protein.

Some methods have been indicated this morning such like PAGE techniques. You may also use affinity chromatography. When a specific ligand for a given protein is known, the affinity of this protein for such a ligand can be taken into account to check its good state of functionality after labelling.

Missirlis: It seems that all speakers so far agree that the need for standardising and knowing which protein you are using is obvious.

Let me put a parenthesis here. I've been working with fixing porcine heart valves with glutaraldehyde. And there have been many papers in the literature where everybody is using glutaraldehyde but none knows what is the glutaraldehyde of the other's. The efforts to standardise that have been hampered because companies are using their own 'secret' formulas, as they say.

So, I hope that this is not the case for these proteins and it seems to me that it would be helpful if one suggests here what standard method is proposed in order to purify or to standardise fibrinogen or albumin or any other protein which is important.

Cazenave: I would like us to discuss the handling of proteins. I have in mind for example coagulation factor VIII, the protein missing in hemophilia A. The way you keep the protein and the way you freeze it is very important. And the way you thaw it is very important because you may denature the protein, so you may have a very nice pure protein and then garbage it because the handling procedure is not good. And this is something which is not in people's mind most of the time.

My second comment would be on differences in animal species. What we are going to say for human proteins, because we have to narrow the subject, might not be true for animal proteins. For example, rat fibrinogen supports rat platelet aggregation but not human platelet aggregation, whereas the converse is true. And this is due to a different length and structure of C-terminus of the γ-chain. So one has to be careful. Proteins are not all the same across the species.

Brash: Just to go back to this idea of quality control monitoring of proteins, I think what you're suggesting is that maybe monitoring should take place not only at the beginning of the experiment but at different stages along the way. And so to my mind we're still on the theme of quality control protocols to check the purity, perhaps at different stages of the experiment.

If we don't know what handling techniques, such as freezing and thawing at different rates are going to do to the protein, then we should check as we go along. That is a good additional point I think.

Missirlis: I think apart from finding out or establishing the need for this, one should propose some concrete ideas: John, may we have your comment on this matter?

Brash: I think I would suggest at the very minimum that all of us should run SDS polyacrylamide gels of the proteins that we use, routinely, before we do experiments with them. This is a standardised, relatively easy and even fast technique.

For example, the Pharmacia Fast system allows you to run a gel now in approximately one hour and check the purity of the protein almost instantaneously.

Hoffman: I have a list of questions concerning the source of the protein, the purification methods, the characterization, storage, handling and then taking it out of storage and characterizing it again as a function of time.

So, should we start and come back later and talk about the sources? There is such a possibility to say we should all use this kind of source. This source of fibrinogen, for instance, if possible, would be the preferred choice. A second choice would be this.

Another question that comes to mind is purification. Then I must ask a rhetorical question: when is a protein pure and when is it not? How much vWF is permitted in a fibrinogen preparation? By what criterion would you say it's pure and so on? These are the questions that come to mind.

Wolf: I would agree with all the things which are told here but on the other side we should consider the goal of our work. It was the goal of our work to look at the hemocompatibility. So, if you start any experiment using plasma or blood then all the properties which we did find out with pure proteins will not work because then the same proteins are in the plasma maybe together with lipids or lipid-protein complexes and so on.

So, for me, it's a question, also, to select really the type of experiment what we want to do. Because you can make at one side purification and on the other side in the more real case all these conditions cannot be fulfilled. Therefore, maybe, we should also discuss the question at what type of experiment is it really necessary to have such highly purified proteins.

Brash: I think that's another question. Possibly it may well be listed in one of the other sessions but for me at least what we seem to be addressing at the moment is: if you decide to do an experiment with protein X, whatever that may be, which may involve adding it back into plasma or blood or studying it as a single purified protein, then you have to know what you're working with. It's as simple as that.

It's not really a question of whether this purified system bears any realistic relationship to blood, or to the condition of the protein as it exists in blood. I think that's another question. Maybe I misunderstand your point.

Cazenave: Should we go along with the list that Dr. Hoffman has provided. I think that might be a good framework and to take one protein as an example. Well, we have to come to some agreement which can be done in any lab.

I think purity of the protein in one way has advantages and disadvantages. What we want to know is what is the state of the protein is and what are the contaminants, rather than having the purest protein, which might not be functional. Can we say the purer we have a protein the less functional it could be?

Brash: Absolute purity is a hypothetical abstraction, anyway. So we should really not be talking about that. But we want to know what is the purity of the protein we have. I would like may be to ask somebody else to add to this list.

Hoffman: The question I'm thinking about is, what's the purpose of all this? Are we to put out a nice publication and who would use it? I think there are really two purposes: One is for new people starting research in this field to get the advice of so-called experts, and then for the experts to try to come to an agreement on what kind of protocol they think would be useful for everybody. This is where I would start from. That would be my goal.

Missirlis: This was also our goal when we tried to work out this workshop. Advice for new comers and also experts coming to an agreement on some fundamental matters.

Cazenave: Should we start from the beginning, I think that your scheme was nice. Maybe Dr. Hoffman would write that for us on the board.

Missirlis: So let's start with sources. I think it would be also important if somebody knows also the cost of certain things to put it down or to indicate it. Sometimes you may want to do something and it's so extremely expensive that it would be difficult too, to carry it out.

Hoffman: Some demonstrating criteria: purity and the type of experiment. Storage and handling obviously and indicated over a period of time. Periodic checking to the continued activity or viability of the protein; so storage and handling gave some feedback here. The period of time is a characterisation, and at the bottom are the functional experiments. The goal is how we use it to illustrate, search out, test hypotheses about protein-surface interactions.

Cazenave: Do we have to start with one protein as an illustration? Should we take fibrinogen, for example or another protein?

Missirlis: Well, which is the mostly used protein among this group?

Brash: I think that's difficult. Frankly, I'm not too keen on the idea of focusing on one protein. Because I think that's distorting the situation. Certainly fibrinogen is used a lot and we could focus on it, but there may be others that should be also considered. And there are certain techniques that you wouldn't want to use with fibrinogen that might be important for other proteins: enzymes, for example. We may want to consider two or three proteins as examples of different protein types.

Cazenave: It's not a bad idea but we have to start. Albumin is a rather blunt protein, I think.

Missirlis: So, let us indicate the sources of fibrinogen.

Cazenave: I'm in a blood bank, so I never buy anything. We've our sources which is rather easy for us, or easier...

Brash: Kabi's fibrinogen has been used a lot by many different labs around the world. This is not to say that it's the best possible fibrinogen available and it does contain several contaminants at fairly low levels. But it's something a lot of people have worked with.

And it's a reliable source. It has been available for at least 15 years and there is no sign that it will become unavailable at anytime in the near future.

Cazenave: ...and it's human.

Brash: Yes it's human.

Baquey: I use to work also with Kabi fibrinogen of human origin. We tried to work with Sigma's but we found a very bad clottability of this quality of fibrinogen; for animal experimentation, fibrinogen is prepared from a pool of plasma collected from several animals.

Cazenave: Do we want to discuss commercial fibrinogen or to prepare and purify fibrinogen? I think Kabi's fibrinogen is a good product. It's prepared by the method of Blomback and Blomback is a real expert on fibrinogen. So the source is quite all right.

There are at least 26 or 30 contaminants and it's very difficult to get rid of some of them. If people want to prepare, they have to be sure, and that will be required, I think, that their donors have no hepatitis or HIV infection (AIDS). I mean there is a danger. You have to have reliable sources and be sure that you don't manipulate contaminated blood. And this is true each time you manipulate blood. Hepatitis is a major problem, but people are not aware, newcomers are not aware. If you work in a lab with human proteins, get vaccinated beforehand.

Beugeling: We also use Kabi as a source of human fibrinogen and then purify it. The purification method we use is the method according to Hamer from the University Hospital Utrecht and it's a personal communication. I think it's not necessary to discuss now how we purify it unless you want to know.

Hoffman: Is it possible for this group to list the appropriate protocols from which a new scientist could pick one or two? One or two that are the preferred protocols for purification; for instance Kabi fibrinogen? Then what further steps do you do? You can come up with a preparation that is useful for your studies. For example, we use baboon fibrinogen which we prepare ourselves from baboon blood freshly drawn, but not everybody has access to this.

So, is the useful product of this discussion to be a kind of list of preferred sources of protein, methods of preparation and protocols already published? They must be in published form.

Brash: If we took Kabi fibrinogen again, this is provided with certain quality control criteria from Kabi. The clottability with thrombin, for example, is usually given as a quality control parameter, and it's usually in the order of 90-95% thrombin clottable.

This indicates, of course, the biological activity of fibrinogen. Beyond that, the product contains a number of salts, which come from the fact that it's rendered soluble by these salts and it's prepared by freeze drying. So, the salts are present in the solid preparation which is provided by Kabi.

This means that you have to reconstitute the material with pure water and then you have to dialyse it into the buffer that you want to work in.

Cazenave: I will add at the appropriate temperature, because fibrinogen is a cryoprotein. So, don't do that at 4°C, as you may have trouble from time to time with precipitation. And I think you are right. There is a lot of citrate and other ions, which have to be known.

Now, as far as protocols are concerned, I think there is Blomback protocol, which was published in a Swedish journal in the 50's, (Ch. 6, Ref. 10) which is still the method used by Kabi and it works extremely well. It's a very good method, but there are other methods too. The one we've been using is the Kekwick method, which is by repeated ether precipitation (Ch. 6, Ref. 11). It gives a very good fibrinogen, but with low yield, about 10-15%, which is rather little for such an abundant protein. And I think about the checks of purity. Well, we are mainly concerned by the presence of fibronectin or von Willebrand factor (vWF). You can remove fibronectin by passing your fibrinogen on an affinity Sepharose-gelatin column, which is a very classical method (Ch. 6, Ref. 12). Most people use this and it works very well.

And for removing vWF it's the same thing. You can pass it through an affinity column on insolubilised vWF antibody. This raises difficulties because you lose a lot of protein, but I think it's important, if you dealing with, let's say, platelet adhesion.

Baszkin: I would suggest that one of the possible ways to proceed would be to start with the commercially available proteins and to suggest a standard method of their purification. This will enable many of the existing laboratories to keep to a standardised protocol.

If one has to procure himself with a protein by extraction from an animal organ of from blood it will draw aside a lot of laboratories working in this area. Maybe one of the experts of this workshop could provide the concerned laboratories with such a protocol.

Beugeling: In fact I want to answer Prof. Cazenave with respect to the purification of the Kabi fibrinogen. Sometimes it's necessary, i.e. Poot found it necessary, to add protease inhibitors in order to prevent degradation of the protein. So that maybe an important point. In his case the fibrinogen with protease inhibitors passed through the aminohexyl-Sepharose 4B column. After elution, vWF and fibronectin stayed on the column. So you remove two proteins in one step.

Cazenave: I think you are right. This is a very common thing and one of the difficulties of starting with proteins from commercial sources. When we purify our fibrinogen we start directly from blood, special blood, not what we use in the fractionation plant. We add from the start a cocktail of inhibitors, and mainly inhibitors of the fibrinolytic system because the effect of plasmin on fibrinogen is one of the key issues. So proteins from commercial sources are not bad, or may be good I would say, but if you want to be sure, you've better to start from the raw material. And this is difficult.

Brash: I would add to the list of things that are being put together regarding Kabi fibrinogen, that a common contaminant is plasminogen. And if you're studying anything at all to do with fibrinolysis, then you must be absolutely sure either that you know how much plasminogen there is in you preparation or that you have removed enough of it, that any study of the fibrinolytic effects is independent of the plasminogen which is in the preparation.

This can be done by passing the original fibrinogen solution through a DEA Sephadex column which also removes fibrinogen polymers that are present in the Kabi preparation.

I might add to this also that if you want to get more quality control data than is normally provided with the sample you can write to Kabi and they will give you a fairly detailed listing of the contaminants which they have shown to be present. Obviously, a company doesn't want to condemn its own product by giving too many bad pieces of information regarding it, but this is something that I have done and they have provided me with quite detailed information.

Missirlis: So you come back to the question that Tom Beugeling raised, that you need a protein for a specific experiment, and this protein may be characterized differently for another type of experiment. But I want to make a question to all of you who have been using proteins for adsorption.

If you start with the same source, say kabi fibrinogen, are there any two labs that use exactly the same procedure? And if not, what are the reasons for?

Cazenave: No, I think there are obviously not two labs who are using the same protein. But I think we are more and more concerned about the quality of the proteins.

What is quite nice in such a group is that we are people of different backgrounds, with different issues in mind. I think people who are doing coagulation or cellular biology, I would say are more aware that a protein is not a protein. It has properties, it has a degree of purity, contaminants and things which will influence its characteristics and fibrinogen is made to be hydrolysed by thrombin, which is also a contaminant. And by plasmin. So, when we deal with cells, we are very critical on those issues.

Well, people involved in chemistry or physical-chemistry are more critical on the quality of the material, the properties of the surface, things we ignore and I think this forum is nice, because we are able to speak to each other and say: well, you have your requirements and we have ours too, and we should put them together in order to have more meaningful results.

And this is very difficult, the problem of proteins. But, I think there are certain contaminants, which in any case, when we are using relatively pure protein should not be there, because they modify the protein you are working with.

Brash: I think the question of adopting uniform procedures in the community of people who do this kind of work, which is what you referred to, is almost impossible. You can think even in terms of type of buffer. For people studying single protein or two protein or three protein systems you can't even get them to agree on what the best buffer is to use.

And you won't convince Prof. X that my buffer is better than his, even though I am convinced that it is. I think the idea here simply is to make what we consider, as a group, to be valid, reasonable recommendations.

Missirlis: Then, can you suggest valid, reasonable recommendations? And what you just said, that even the buffer cannot be agreed upon, it shows something. What is it?

Cazenave: At least we can ask the question. I think this is important. We cannot say: you should do that. I will give you an example of buffer: imidazole has been used a lot as a buffer. Well, if you are dealing with platelets don't use imidazole, because imidazole is a thromboxane synthetase inhibitor and is now a source for drugs. We didn't know that before, but because people think that buffer is not relevant, so any buffer will do. I mean the quality control, 'it works, it's good', is wrong. 'It works', I want to know why it works, and this may have importance. As long as the question is there, some people will answer it differently in different settings.

Baszkin: I would like to come back to my proposal. The only possible solution of this problem is to suggest a simplified protocol of purification of commercially available proteins.

It is not possible to do it, otherwise. A simplified protocol of purification of a commercially available protein of interest should be sent to the chairman of this workshop and be distributed among the different groups, involved in the protein adsorption work.

Brash: Well, I think that's a noble long-term goal, but I would challenge you to come up with a list and a finite number of products that you would be prepared to provide this kind of information for.

Hoffman: Let's try and establish a couple of criteria. For fibrinogen, let's say we have a source or two that we can recommend and, in terms of method of purification, a buffer, the particular quality of purity.

We've mentioned, for example, impurities such as plasminogen. There is also vWF. Are other impurities that we must remove fibronectin, vitronectin? To talk to Deen Motion now thrombin is much more important than fibronectin.

We have now at least some proteins to remove. Can we identify one or two protocols for their removal? For example, if you are doing platelet adhesion, you have to get vWF out, and so on.

I'm not going to go further, but this is just to stimulate the next question. Can we make a list of those critical properties that we want to remove?

Baquey: We must consider also the particular use of the protein. If the protein is aimed to be used in an experiment as a major component of the protein solution involved in the study, it is quite important that the protein conforms to the expected specifications of purity.

But if the protein is just used to be labelled, for instance, and add as a tracer, it's important to check that this protein has the right functionality. For example the thrombin present with Kabi fibrinogen, which is added to plasma in our experiments, is, I think, of no matter, because the thrombin will be at a very low concentration. But if there is unexpectedly thrombin in a solution of fibrinogen 2 mg/ml, it will be of great importance. I think it's quite different if you use the protein as a tracer or if you use the protein as a principal factor in the system.

Wolf: There was mentioned that there will be some primary reference biomaterials distributed over different laboratories. So, it's my question, if one or some of these proteins could be prepared in one of the laboratories, where the greatest experts are working in that field, and to distribute them in a purified form to the other labs. Or, that is a more practical question, would it be possible to engage a company which has a great experience to make some selected types of purified proteins, which can be used by interested groups?

Because I think that the efforts to make the same purification procedures in the different groups are relatively different, because of knowledge about all the details.

Missirlis: So, there is a suggestion. But how about a company or a blood bank or something?

Baquey: We had a small example five or six years ago. We organised a group in France about hemocompatible materials and we concluded a contract with the blood bank in Lille (France) as a supplier for a special quality of AT III.

And everybody working in the group was working with the same batch. That may be less easy for an international group, but we could have some discussion about that.

Brash: The fact that you mentioned antithrombin III seems to me to suggest that we should be making a distinction between abundant proteins, like fibrinogen, and proteins that are relatively scarce, such as antithrombin III, and which are not commercially available.

And it may well be that Prof. Wolf's suggestion would be more appropriate as a subset where the facilities and expertise of a blood bank might be required to prepare proteins like AT III or vWF or HMWK, which, obviously, are not going to be available in quantity from commercial sources.

Cazenave: No, because for commercial sources there is no market.

I think I would see three sources. The best source is human blood and you make your own proteins, especially if they are scarce proteins or if you want to control all the steps from bleeding the donor to the end of the process of purification. This is still possible and many labs do it or will do it, when they want to be sure of the quality of their proteins.

For more abundant proteins, like fibrinogen, or albumin, or vWF, which is not abundant, but which is in cryoprecipitates, well, you can get them from commercial sources. These are products, which, most of the time, have been inactivated, but still the protein is there, I don't think it's too denatured, and you can purify from the source.

Then, some of the products are available from other companies, for example, Kabi is widely distributed. Sigma has some products of quality, but it's very expensive when it's very pure, so you can achieve that at a lower cost, I think, and be sure of the quality of your products.

In different countries, and I would go along with Charles, let's say in France, it's not a major problem to get proteins from the blood-banking system. In Holland it's the same, I think, you have access to pure proteins or to the raw material, which you can purify. This is not difficult.

For example, we now have a fibrinogen, which is safe, virus inactivated, 95% clottable and 92-98% pure. And which can be used in humans. So, it's not difficult to have access to that, and eventually to repurify it to remove fibronectin and vWF which it contains, but, for human use, it's a useful contaminant, I would say.

Missirlis: So, after one hour of discussion on this problem, Alan, do we have something on which most people have agreed?

Hoffman: Basically there seems to be one good source of human fibrinogen. I am not sure about other sources. In terms of protocol of purification, I'm still not clear on exactly what you would tell a young investigator.

You see, if we wanted we could list all of the impurities in decreasing order of importance from what we know today. But where do we draw a line? Which is the most important impurity to make sure it's out of a preparation?

Because, why do we want fibrinogen? Well, for many reasons. Some people use it for platelet receptor interactions, others use it for simple criteria for establishing the effect of various surface treatments. It's supposedly good if it does adsorb to a great extent.

So, you have different uses, and if you have impurities in the fibrinogen preparation, it really doesn't matter, it seems to me, for using it as a probe of surface chemistry.

If, however, you're going further and looking at cellular interactions, such as platelets, then you must worry tremendously about these impurities. If you are looking at the thrombogenesis on a surface, independent of platelets and platelet aggregates, then you also have to worry about certain impurities.

So, what I was saying is, if I were a new investigator in the field, I would not be comfortable yet with what has come out.

Missirlis: I'm in your position. Willy, do you want to say something about this? You said you had some experience with protein adsorption and then you did not continue. Was it because you couldn't find purified proteins?

Lemm: No... it's not so easy and the outcome was not really sufficient, so we dropped the whole thing. For example, I can mention that we used a smooth silicon coating for artificial hearts. But it had a high preference for fibrinogen. And we could not explain this effect. So, the affinity to fibrinogen was high, but the non-thrombogenic property in vivo, in an artificial heart was excellent.

Cazenave: There is a simple explanation. You have to present to platelets, for example, the right domain of interaction, as shown by using antibodies to detect fibrinogen on a surface. If you don't fix it in the right conformation, you may have a lot of fibrinogen, but it's thromboresistant, to a certain extent, towards platelets.

Brash: Thromboneutral, maybe. I'm less pessimistic than Alan, as to what we have been able to discuss and come to agreement on. It seems to me that his listing is one outcome. Also what we have really said is that somebody working with proteins should identify the source, should be aware that perhaps additional purification may be necessary, depending on what he wants to do with the protein, that he then has to do a minimum of characterization on the protein to verify that his purification technique has done something and that at least one of those techniques should be SDS PAGE.

Actually I would add that some kind of immunochemical verification should be done, especially in the case where laboratories are isolating trace proteins by their own techniques beginning with blood or plasma.

Because, of course, SDS PAGE is only an indication, at best, of the number of protein contaminants you have and their molecular weights. There are many proteins that have bands in the same molecular weight position. So, the only way that one can be absolutely sure is to do an immunoassay of some kind.

Cazenave: Would you say reduced and non-reduced SDS-PAGE systematically?

Brash: Both, maybe.

Baquey: And affinity chromatography may be added, too.

Brash: It's not always possible to do that.

Baquey: I mean when it is possible.

Missirlis: So, this may be difficult for a young investigator.

Baquey: Maybe yes, but, I come back to AT III, heparin Sepharose® is an affinity phase which is commercially available.

Cazenave: Again this is incidental, but if you do affinity chromatography on heparin-Sepharose to recognise that you have AT III, you select the molecules which are reactive for this type of heparin, so, again, you have difficulties.

It's one criteria. I think the simple one is SDS-PAGE reduced and non-reduced. Then, what type of immunoelectrophoresis will you do against it, an anti-human total serum or a specific antibody?

Brash: I guess either way. Possibly the simplest way to do it is by Ouchterlony double diffusion. It's fairly unambiguous and it's easy.

Cazenave: With a precipitating antibody.

Brash: Yes.

Missirlis: How about the last item, storage and handling?

Brash: It seems to me that we've already said that this characterization procedure should be applied along the way following storage and handling.

Now, as to what's good and what's bad in storage and handling, that's something we have not discussed.

Missirlis: Can you suggest what is the best way to store and handle the protein that you have purified and initially characterized?

And then the idea, I think, of Prof. Hoffman was that when you are ready to use it you have to characterize it again.

Brash: Yes.

Hoffman: May I make just another comment? Most people are using radiolabelling and we didn't include this on the list. That is really the step that you need to do just before you do your experiment, your protein adsorption. That has to be added to this list.

And actually, I'm not that pessimistic. We have made some very strong suggestions of what you have to worry about. To me the next step would be to provide, if it's possible, a recommended list of references for the new researcher.

These protocols would be trustworthy for these authors. These protocols for storage and handling are also recommended. This one for radiolabelling is the best. Is that possible?

Brash: I think perhaps a list of references could be provided to a new comer who might not be familiar with SDS PAGE and one could recommend a reference (e.g. Laemmli, U.K., *Nature*, 227, 680, 1970).

Hoffman: The NIH, several years ago put together a group of experts to recommend a series of blood compatibility tests. What they ended up with was a series of tests to be recommended to be done, some of them in sequence or at least in groupings, and they gave references of where the protocols could be found in many cases. I think that would be very useful.

Missirlis: Of course somebody must do it. And, yes, one could indicate that everybody who talked now could suggest some references, even if they don't agree with each other.

Brash: There can be more than one reference.

Hoffman: I think when the tape comes back, then, everybody will add whatever they can.

Missirlis: That's a very good suggestion.

Cazenave: If we say a few words about storage and handling. I think one of the difficulties is that many times people want to freeze their protein solution and then thaw it. And I think there is a lot of damage which is done at that time.

Well, with albumin it's alright, but many proteins are very difficult. And it seems that the fastest way of freezing proteins, either in vapour or in liquid, nitrogen, is the best. Some people have even told me that they are dealing with purified hemoglobin, they make pills by dropping droplets of protein solution into liquid nitrogen. Then it's very easy to handle.

We don't do that, usually. We freeze small plastic tubes in liquid nitrogen and it's good enough. Then we keep them, depending on the protein, at –30°C or –80°C.

And thawing, proteins have to be thawed in water bath most of the time at 37°C, then eye-control and when it is done you put them on ice. But if you do that after letting them thaw in room temperature you have a lot of problems. I'm not a physical chemist, but I've been told that going slowly through the eutectic point is very bad for proteins. This is true at least for factor VIII. If you do that you lose 20% of activity.

So, I think you have to be careful. Many people and many chemists add sodium azide in their column and to their protein. Newcomers have to know that azide can be dangerous, if you put it in your sink, it may blast. And sodium azide is one of the best platelet function inhibitors, working at 10^{-9} M concentration. I recall, because I've gone through that, when I was working with platelets and C1q in Toronto. The biochemist was collecting C1q with sodium azide and we had one of the best inhibitors of platelet function. Fortunately, he gave me the buffer with sodium azide as a control and it was a very good platelet function inhibitor. So, if you work with cells, it's a very dangerous poison of cell function. Don't use sodium azide to avoid bacterial proliferation.

Brash: Sodium azide is possibly the 'standard' substance which is used to control bacterial growth in protein solutions. And so, many people do in fact add azide to their buffers and their storage media in general.

Cazenave: Yes, that's true. But you can rinse your columns before doing that and then, if you want to store your proteins for a long period of time, you can filter them with a Millipore filter. Well, you may lose some of your protein, so you have to be aware of that. But you can keep proteins without sodium azide when they are frozen at –80°C. Bacteria don't grow. On the other hand, if it's to work on surface interactions with no cells, azide is all right.

Missirlis: So, thawing the frozen protein from –80°C to, say, + 22°C does not effect the functionality of the proteins, or does it depend on the rate of thawing?

Cazenave: It depends on the rate of thawing. You have to do that quickly in a water bath at 37°C and when the protein is in solution keep it on melting ice. If it takes two hours to thaw, it might be very deleterious for the protein.

Hoffman: This is where you can use a fairly simple criterion. If you thaw the protein too fast and you check on PAGE and find fragments you don't want, then go back and modify the thawing procedure. This is something even a new investigator can play with.

Cazenave: It's sometimes also the function. You don't have fragments. It's the biological activity, factor VIII, for example, is very critical. I mean you lose activity but you don't have fragments. And freezing and thawing repeatedly is usually bad, so you have to aliquot your proteins.

Brash: I think that we all agree that freezing and thawing is something that we have to be careful about and pay attention to. And standardize, possibly, and check, if we've done anything bad. Another handling problem...

Question: Would you say what you would check and standardize?

Brash: Well, obviously it depends on the protein in question. I think SDS-PAGE is fine. That will indicate primarily fragmentation of the protein into smaller pieces, if that happens, which is a fairly drastic effect. It will not detect denaturation of any kind, so I think that, as Jean-Pierre says, if you are working with a protein of which the biological function is the critical thing, then you must have a test of the biological function, following freezing and thawing.

And something else that one may want to beware of is induction of foaming during handling of protein solution. It's the most natural thing in the world to have a solution in a test tube or a flask and shake it to aid in dissolving. One simply shouldn't do that with protein solutions, because they are surface active molecules, they are detergents, they are going to form foams and you are going to, therefore, expose your protein to a solution-air interface, whatever that may do, and in general it does bad things. That's a fairly simple point, but, maybe, one which might escape the novice.

Missirlis: Can I ask a very simple question? How long does it take for someone who gets fibrinogen from a company until he freezes it? I mean going all through these methods is it something that takes 24 hours or a few hours?

Cazenave: To purify from blood? One week.

Missirlis: Not from blood, from a commercial source.

Brash: Kabi fibrinogen comes in the form of a freeze-dried solid preparation and you can take it from that by dialysis into a buffer possibly over night. You then aliquot the dialysed protein into small portions and freeze away. This can certainly be done in a day.

Missirlis: O.K. One other simple question. The flasks where you keep this, either glass or plastic, does it affect the protein that you want to purify and have it functioning for your specific experiments? Is it something that concerns people?

Brash: The material of the container does concern people. It concerns biologists, I think, more than possibly it concerns a physical chemist. And, for example, the conventional wisdom among coagulationists is that it's better to store in 'plastic' than it is to store in glass, because less protein adsorption will take place to the plastic than to the glass, which is of course completely wrong. That certainly is one point.

I think it depends again on the proteins that you are storing, what their function is, and what you intend to do with them afterwards. Glass, for example, is a more active, more dynamic surface than most hydrophobic polymers like PE, PS materials which are likely to be used to make containers.

I think mostly what happens on those kinds of surfaces is that you get monolayer adsorption and that's all. They are not that dynamic. So, maybe, it's better to use plastic containers from that point of view.

Missirlis: Any other comment before the break?

Baszkin: I have one. As far as surface chemistry is concerned, it is important to know what impurities contains a protein to be used. The adsorbed quantities at a given material vary according to the purity of the protein, and the data of the same material, obtained by the same technique, but coming from different laboratories, may be different because protein purity was not the same.

13. Labelling and protein identification techniques

Missirlis: We have four items to discuss now: labelling and identification techniques – how important they are in the protein identification? – which blood proteins are important to be looked on – we have touched on it already – and isolation and purification methods for blood proteins.

Already the backround of this has been set and hopefully some more concrete statements will be heard this afternoon. So, who wants the floor?

Baquey: Maybe I could try to start the discussion. If we look at the labelling methods which are used for proteins, most of them are based on radioiodination and, all of them, suppose a transformation of the commercially available isotope source from the state of iodine ions to the state of forms having a higher degree of oxidation. And these iodine forms are able to bind to the proteins; but in a second stage the excess of oxidative species must be reduced in order to stop the reaction by adding a reducing agent.

So, you must pay attention to the effect of the reducing agent on the protein. And proteins which bear disulphide bonds may be fragile upon such an action. So, the protocol of labelling must be designed in order to avoid any deleterious effect as well of the oxidizing agent used in some protocols as the deleterious effect of the reducing agent which is often used in several protocols.

Lemm: Pr. Baszkin you said this morning you use ^{14}C. How do you introduce it?

Baszkin: Well, we have two methods: acylation (Ch. 2, Ref. 4) or methylation (Ch. 2, Ref. 5).

Two checks are important before you perform an adsorption experiment from the solution containing a single protein. You need to know if the diffusivities of the labelled and unlabelled proteins towards the interface are the same. This can be checked when you plot the protein solution surface tension versus its solution concentration. If you have the same curve this means that the labeled and unlabelled protein molecules arrive to the interface with the same speed. This means that the diffusivity of your proteins is the same.

Y.F. Missirlis and W. Lemm (eds), Modern Aspects of Protein Adsorption on Biomaterials, 113-135.
© 1991 *Kluwer Academic Publishers. Printed in the Netherlands.*

Now, if you put a polymer film at the top of the protein solution there might be a different uptake of the labelled and unlabelled protein by a given surface. You have to make an isotopic dilution of your hot protein with different amounts of the cold protein and check if there is any preferential adsorption of one of these proteins on a given surface.

If both checks are positive it would mean that your labelled protein has its surface properties unchanged. I'm of course not speaking about the possible change in biological properties of a protein, which may occur during the labelling procedure.

Wolf: May I ask you the following question? In this colloid chemical investigations the molecular weight can be considered as the main parameter. If you change the conformation of your protein by this procedure of labelling, do you think that you can measure it with a Wilhelmy method?

Baszkin: You can take another method of measuring surface tension. You can also measure the surface viscosity instead of the surface tension. If there is a change in the structure and the conformation of your protein this would appear on your curve. That's what we believe. But, maybe, you can suggest another method.

Wolf: I cannot suggest another method, but surface tension is a thermo-dynamical parameter, which is not coming from the statistical thermo-dynamics, that means, from the equilibrium thermodynamics. That means also if there is only a change of the secondary or tertiary structure of such a molecule which could be done by labelling procedure, it could in my opinion not be measured by such a method. Only if there would be a real change of the molecular weight of the molecule, or to break down really a lot of charge groups or so that would be possible.

Then you could detect, with the proposed methods, such a conformational change of the molecule. I cannot offer another method, but, I think, from these parameters you cannot get this information. But if there is a break down of some molecular chains or something, then it should be detected.

Baszkin: Well, I definitely do not agree with you, because the values we use in our plots are the values of the surface tension taken at equilibrium. This means that each point of this curve is the surface tension measured after 20 hours at plateau. This is not just a surface tension instantaneously measured.

Brash: Adam, you said at the beginning of this presentation that you use this as a measure of the diffusivity and the integrity of diffusivity of the protein after labelling. But you measure surface tension after 24 hours. And so I have some difficulty understanding the connection. If you've reached thermodynamic equilibrium at that point, diffusion should no longer be in play.

Baszkin: You have also the plots of surface tension as a function of time, because the surface tension is automatically recorded on a chart as a function of time. There is always a certain time which is necessary to bring all possible molecules to the interface. If, for the labelled and unlabelled molecules, at a given protein concentration, this time is the same what this would mean? This means that both proteins are adsorbing with the same rate.

Wolf: It's correct, but you get not any answer about the change in structure. What I want to say is, if you have, say, isoenzymes, fully different but as a molecule they are equal, and you find by these measurements that they are equal in that physical property the diffusivity of that protein molecule depends on the molecular weight and the size of the molecule.

If the molecule is a little bit unfolded or changed I think you cannot get from this thermodynamical parameter that information that you want to get.

Baszkin: Well, the question is the following: do we change the nature of the protein when we label? That's the main question. This can be answered from different points of view. As far as adsorption experiments per se are considered, I believe that if you perform these two checks and you have a positive answer there is no difference, well, at least, you can say the surface properties of my protein are not being changed during the labelling procedure. Biological properties, that's another question.

Cazenave: Do you have positive results meaning you take one protein and you label it with different amounts of, let's say, ^{125}I and different times of incubation and if you reach drastic conditions you see a difference? Because there is a problem of the sensitivity of your method. Maybe this type of measurement is only able to pick up, well, drastic changes.

Baszkin: There are some proteins that we are able to label without any problems. To give you an example, we can label albumin without any problems. When I say without any problems, I always refer to this type of experiments.

This means labelled-unlabelled proteins give the same results of adsorption to an inert surface, which is for example a PE, or the solution-air interface. There are also proteins for which we do not obtain the same results. For instance we have some problems with fibrinogen and we don't know which is the source of the error.

Hoffman: Actually, what we are doing here is reverting to a discussion of methods of characterization, which is of value, but at the same time everybody has his favoured 'pet' method. I would be a little uneasy with the surface tension method, because it involves air surfaces as a method to test the purity of a protein.

I still think it's interesting to talk about your reasons for using ^{14}C and tritium. And I think you mentioned it's mainly because you don't want to pick up the signal from the bulk solution. You only want that which is near your test surface. Is that right, Adam?

Baszkin: The main reason to use ^{14}C is that you can measure adsorption in the in situ conditions and that you can distinguish between the loosely and tightly bound proteins.

Hoffman: May I add a question? Have you ever established the validity of your labelling techniques by comparing the iodine label with your ^{14}C label?

Baszkin: Yes, but it is very difficult to compare them. Because with the iodine method before you measure you have to rinse.

Wolf: And what with your first methods, your surface chemical methods? Did you find any difference between iodine labelled and ^{14}C labelled proteins?

Baszkin: Of course. I worked with iodine labelled proteins but not necessarily on the same surfaces as with ^{14}C labelled proteins. We have never done a systematic study to take the same surface, to label the same protein by different methods, and to adsorb it in the same conditions. So, it is very difficult to answer such a question.

I would like to point it out once again. The iodine method does not permit to know how much of the loosely bound protein you have on a given surface.

Lemm: Is this a reason, that ^{14}C is a pure β-radiator while iodine is γ-radiator? Is this a reason?

Baszkin: The reason is that the solution containing a ^{14}C protein attenuates completely the radioactivity which originates from the bulk. This means that what you measure comes more than 50% from the adsorbed protein molecules at the interface with the polymer and to a certain extent, depending on concentration on which you work, from a very thin, adjacent to the polymer solution layer.

When you measure adsorption at the polymer-protein solution interface with γ-radiator you count the radioactivity coming from all your solution and you cannot discriminate the bulk radioactivity.

Brash: I just want to point out that I appreciate the fact that this is an in situ technique and has considerable merit for this reason, but I would disagree with your statement that you cannot measure in situ with iodine labelling. There is a way. I could give a short seminar on it, but I don't think it's appropriate to do this at this point.

The method I'm referring to has been published already by a group in France in Strasbourg, which you undoubtedly know about. And we have begun to use that method ourselves (see Ch. 17). The surface must be in the form of particles, or, at least, there must be a sufficiently high surface-to-volume ratio that you partially deplete the protein in the solution.

Baszkin: Then, if you have a material with a high specific surface, then you are confronted with another problem. You don't know what you have on the surface.

Brash: You can measure the surface area of particles.

Baszkin: But you don't know what functional groups you have on such a surface.

Hoffman: Excuse me, but we can use ESCA to characterize surfaces at least. That technique would tell us.

Baszkin: On powders, on small particles?

Hoffman: Sure. We have done it.

Jennissen: Did you label your protein by adding a tracer of labelled protein to unlabelled protein or did you add the tracer protein to the cold-labelled protein?

Baszkin: We label our proteins by putting the label on a protein and we dilute isotopically the labelled protein with a cold protein.

Jennissen: Is the chemical nature of the protein the same then?

Baszkin: It is a difficult question. I repeat: we perform an experiment in which we measure surface tension of a protein as a function of its concentration and we take the points at the equilibrium. If we don't find a difference between the labelled and unlabelled protein, we think that they have the same surface properties.

Jennissen: This means that you do not cold-label your protein. That is you do not control-iodinate the rest of your protein with non-radioactive iodine.

Baszkin: No.

Brash: I have disagreed on the same point before. I simply don't see the logic that says that when you look at 24 hours you're measuring anything at all to do with diffusivity, because at equilibrium the system is no longer diffusion-limited.

Perhaps, if you would follow the surface tension as a function of time, then, I'm convinced that you are doing something valid. You showed a curve of surface tension versus concentration at 24 hours.

Baszkin: We plot the surface tension as a function of time for a given protein concentration before it is labelled. And we take the points at the plateau values of such a curve. We do the same with a labelled protein and we take the points at the same time at the plateau. But you can compare points at any time, not necessarily at plateau, because, I repeat, the surface tension versus time is continuously recorded.

Brash: Are the surface tension versus time curves for labelled and unlabelled protein superposable?

Baszkin: If you did not change by your labelling procedure the surface properties of the protein you have the same curves. If you change these properties you do not have the same curve. If the curve is not the same something has happened during the labelling. Now, what has happened that's another question.

Brash: I think your second validation technique, which I also use, is...

Baszkin: But, John, this is not the same thing. When you adsorb proteins onto a polymer then one may ask the question if the uptake of the cold and hot proteins on the surface is the same.

If you have a polymer which bears on its surface two or more different functional groups, then one of these groups may be more sensitive to bind a labelled protein than another one. The same holds true for the unlabelled protein. So this experiment gives us not the same information as the previous one.

Cazenave: Using labelling techniques what kind of isotopes are we talking about? I think there are two ways of study. We work on purified proteins and we want to incorporate β-emitters, is there a problem of specific activity, that is, does tritium induce radiolysis of your protein or not?

Or we want to have a γ-emmiter to be able to do eventually γ-camera imaging, should we use iodine?

What is the problem of the half life of the isotope? We use technetium or indium with cellular proteins. There are other ways of labelling proteins which are very difficult, but which can be used in animals. One way is to pulse label the animal with aminoacid precursors, so we don't do bad things to the proteins, we have native labelled proteins and we can use them. I think that might be the question I would like to address to have some ideas.

Missirlis: Would you please continue with your suggestion of what one should do out of all these possibilities?

Cazenave: Well, I think the protein labels we've been using most of the time are gamma emmiters like iodine, indium or technicium because they are easy to handle, we can count on surfaces which are of different shapes and the radiation goes through the test tube. I think that's one of the reasons why γ-labelling is so useful and popular.

The problem of iodine, as I think Charles has already stated, is that usually you need, if I'm right, to have tyrosine residues in the protein. If we do labelling studies in a very harsh way you can overlabel residues and have damage. And one has to recall, for example, for growth factor receptors or for growth factor labelling with iodine usually changes completely the biological activity of the growth factor and you detect fewer receptors on the cell surfaces. That may happen with protein and polymer surfaces.

Maybe, well, the specialists of labelling can go on and say that there are different techniques of labelling with iodine. There is the Iodogen technique, which we have learned from the Netherlands, which is rather gentle, I think, and there is the lactoperoxidase, another technique (Ch. 6, Ref. 13). Maybe, this is what people want to discuss now.

Jennissen: Yes, I think we should discuss the iodination technique. If you use the chloramine T technique you often oxidize methionine residues and this changes the protein function considerably.

In the case of calmodulin, for example (Ch. 8, Ref. 25), iodine labelling by the chloramine T method abolishes the biological activity, so that it will not activate enzymes any more. Although on an SDS gel or by other criteria you detect no change, the function has been totally altered.

Hoffman: Tom Horbett developed a technique based on the McFarlane ICL technique back in 1959 and I'll put that reference in the text. A number of people are working with PEO surfaces and PEO is known to sequester iodide ions. Therefore, we add to the citrated PBS buffer, which also has some azide, a small amount of cold sodium iodide to swamp out that effect.

Brash: I think this also raises another point about iodine labelling, and perhaps labelling in general, which is that, again, as with the purity of the protein, one should check for loss of label at an appropriate time in relation to the beginning of the experiment.

Iodine labels are notoriously unstable and can be lost from the protein at a rather alarming rate, depending on the conditions. As much as tens of percent per day can be lost if you simply leave the protein at room temperature. We try to avoid this problem by labelling with iodine on the same day that we do the experiment.

We also measure the percent of unbound iodide in the preparation we are about to use in the experiment. This is a very simple thing to do.

Hoffman: You dialyse it.

Brash: No, you precipitate an aliquot of the protein with trichloracetic acid and measure the radioactivity in the supernatant and in the precipitate.

This is not the same thing as efficiency of labelling, which you would measure immediately after having done the labelling procedure. Following the measurement of efficiency we do a separation of unbound from bound label. There is still a small quantity of unbound iodide left after the separation. It's inevitable. You should know how much it is.

This relates to your point regarding the uptake of free iodide during adsorption too, so I think it's important to do the labelling as close as possible in time to doing the experiment with the labelled protein, if it's iodine; ^{14}C I don't know, because I have not worked with it. I imagine it's more stable, but perhaps not.

Cazenave: A small comment on what we have encountered with cells and I would think it's the same with surfaces. Each time you count radioactivity on a surface, well, in logical terms, what you have is radioactivity, it's not a protein. You should make sure by eluting that it's bound protein and it's not a fragment or free isotope which you are counting. Most of the time we don't do that, but cell biologists require that when you define a receptor associated with a cell you must be sure it's still the whole molecule which is there. And, well, isotopes are nice for a non-chemist, I love them.

Jennissen: I would just like to call the attention to the method (Ch. 8, Ref. 3, 14) we employ for the quantitation of bound/unbound protein species. We do not employ a blood protein but an enzyme, phosphorylase *b*. Phosphorylase contains a pyridoxal phosphate coupled via a Shiff base to lysine. In this case you only have to reduce this Shiff base with either tritium or hydrogen.

Thus cold labelling here is possible without changing practically any facet of the structure relative to the labelled species. Thus labelled and unlabelled enzymes are structurally identical.

Through the reduction step the specific catalytic activity goes down from 80 to 50 units/mg, but this is the case for both the 'hot' and the 'cold' enzyme. If you could find a similar labelling method for blood proteins it would be ideal.

Cazenave: Is it the periodiate tritiated borohydride or the pyridoxal borohydride method you are using?

Jennissen: We employ tritiated sodium borohydrate.

Cazenave: Yes, that can be done with proteins, but you have a problem in that most of them are glycoproteins, so you may change the function of your protein by changing the sialic acid which is coupled.

Jennissen: Yes, that is true, but in the case of phosphorylase that is not the case since it is not glycoprotein.

Baszkin: Yes, I still persist; I do not intend to convince you to any type of radiolabelling, but I think that if you don't have an experimental access to the adsorbed and desorbed quantities for different protein-polymer systems your information is not complete.

There are polymers on which protein is only loosely bound, there are proteins and polymers where loosely bound protein amounts up to 80% of the total adsorption and there are systems where all adsorbed protein is desorbable.

This is extremely important since the process of exchange between the adsorbed protein and the protein in solution may occur as well with the loosely bound portion as with that portion which we call irreversibly bound portion.

Brash: I agree that one of the objectives of people who study protein adsorption should be to try to develop more techniques which are in situ techniques and which don't require the separation of the solid adsorbing material from the solution before the measurement can be made.

At the present time there are very few such techniques. There is your technique, Adam, there is the iodine labelling technique using particles, there is the FTIR-ATR technique and other optical techniques based on total internal reflection (see, for example, Andrade, J.D. (ed.), *Surface and Interfacial Aspects of Biomedical Polymers: Vol. 2, Protein Adsorption*, Plenum, N.Y., 1985).

Most of these optical techniques are not calibratable, but I want to agree with your point, that in situ techniques are something that we should be aiming for more and more in the future. Unfortunately we don't have very many good ones at the present moment.

Missirlis: But which are the good ones at the present moment? The ones you just mentioned?

Brash: These are the only ones that I am aware of.

Baszkin: But the best technique is that which enables to measure adsorption of unlabelled proteins. Since you do not have problems of labelling and you can work with proteins which were not submitted to any chemical reaction.

Baquey: I think that you may use isotopic techniques to study in situ adsorption: it is a question of volumic activity and it is a question of double labelling, because if you count, as we do in our ex vivo experiments, and as you do in your experiment, at the same time the bulky solution and the material which is deposited on the surface you must determine specifically the contribution from the surface. This contribution must represent a significant proportion of the whole radioactivity. I mean that, if you count a total amount of 10000 counts in a given period of time, which must be as short as possible to get a good precision on the kinetics, the contribution of the wall must be larger than 200 counts, because it must be larger than the statistic fluctuation.

So, if you have a good ratio between the surface and the volume, or between the surface and the radius for a tubing, you may get a very good sensitivity in measuring the amount of protein which is deposited onto the wall. For example, for a 3 mm tube (Fig. 1) a 15 μg cm^{-2} concentration can be accessed. It is still very high, but if the diameter of the tube is decreased down to 1.5 mm, the lower level of the amount which can be detected on the wall is decreased considerably.

So, most of the time a second label to take into account the real volume you are counting is needed. I mean a label which is related to a species which is not interacting with the wall.

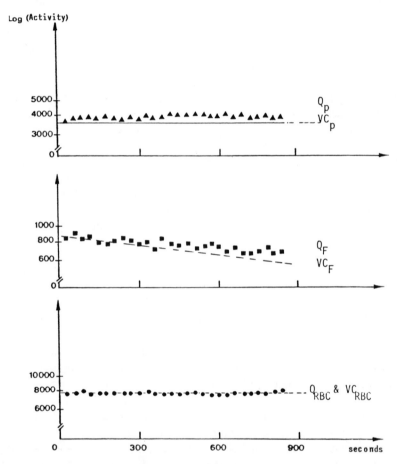

Fig. 1. Recorded radioactivities for the shunt and related to each of the involved tracers during extracorporeal circulation of blood.

Baszkin: I agree with you. You must also have the amount of the protein, or the amount of the biological spacing which is consumed by the surface must be negligible compared to the whole amount of the protein involved in the experiment. Under this condition you may consider that the concentration of the spacing is constant in the bulk, and, as you have the volume, you must deduce, from the difference between the count in the bulk and the count in this part you are looking at, you may deduce the part coming from the wall.

Wolf: I want again to make a comment. I think most of the presented and discussed methods are very good methods and the ^{14}C method has the great advantage that we should not make this rinsing procedure.

If I remember correct, then, it was shown by John Brash that we have an exchange of labelled and unlabelled protein after adsorption at the surface. So, my comment is the following: maybe you can make an experiment with ^{14}C labelled HSA and iodine or other isotope labelled HSA and repeat this experiment under equilibrium conditions with the same reference surface. Then you should find the same exchange of the protein if the proteins are not changed during the labelling procedure and they should behave equal.

Brash: There is no question of the validity of the ^{14}C method.

Wolf: The question is how the molecules are changed during labelling procedure; and if they are not changed or equal during the different labelling procedures, they should show the same exchange with any type of labelling.

Brash: My comment was in respect to Charles Baquey. I think this is a great idea. We have tried this double labelling approach, where you use a label for the bulk fluid, which doesn't adsorb and gives you a bulk marker, and another label for the protein which is both in the bulk and on the surface. We find that typically you are subtracting 99.9 from 99.999 and that, at the counting levels that one can reasonably achieve, it's very difficult to get reliable data, data that one can really feel confident in.

Now, maybe we are doing it wrong and I think it would be extremely important, as a part of the outcome of this Workshop, for Charles to give a detailed description of how they do it, because we have failed to do this.

To make this method work, I believe, requires a high surface to volume ratio. If one is working with tubing which is geometrically suitable for such experiments, it may well mean going to 0.5 mm diameter or less. There are only a limited number of materials available in such small diameters and coating of polymers from solution to such small tubes is difficult.

So, there is that difficulty. But I like the approach although personally I have not been able to make it work.

Baquey: I think there is another method which can combine the method that I described and the method of Adam Baszkin. It is a method based on the use of plastic scintillators; you may drill a channel through a scintillator block and you may coat the lumen with a polymer, and using beta-emitters you may have good performances, better than with a gas counter, because you have a 4π-geometry.

Under these conditions, if tritium, which is the most suitable, or even ^{14}C are used, you count only the radioactive material on the wall and in the first layer of the bulk which is what you are interested in.

Baszkin: Well, there is still another advantage, to have the backround. You can have a direct access to the backround by extrapolation of your curve that you register, which is counting as a function of time to time zero. At time zero there is no adsorption, so counts at time zero correspond directly to the backround, automatically, for a given concentration.

Missirlis: Are there any other identification techniques? Apart from labelling with radioisotopes?

Magnani: By considering the problem of the labelling, I think FT-IR spectroscopy is a useful technique to identify a protein at a surface. Proteins can be characterised on the basis of their intrinsic IR spectra without the need for extrinsic labels such as radioactive tags. In this case the eventual conformational changes, due to the introduction of foreign groups in the protein structure, are avoided.

Missirlis: Is that what Adam wanted to have?

Baszkin: This is my dream, but there are some technical problems. I was very amazed visiting Neuman's Laboratory in Toronto, who performs a very simple experiment. He is just trying to weight protein adsorption with a microbalance. He puts a plate inside the protein solution and he weights the increase in weight due to the adsorption as a function of time.

Of course, by this technique you can only have the data concerning adsorption of a protein from a single protein system.

Jennissen: Another method for monitoring protein in depletion experiments (Ch. 8, Ref. 3) is to make use of the enzymatic activity. In this case you can control the function of your protein by measuring the specific catalytic activity before and after the experiment. This data yields information on possible conformational changes or on the denaturation of your protein during the experiment.

Brash: Yes, these equilibrium depletion methods are classical and have been used for many years, but, as Charles, I think, said, you can only measure at equilibrium, or perhaps at one point on the kinetics curve. It is rather difficult to follow kinetics with this kind of approach.

Jennissen: No, that is not true in such a generalization. The measurement of kinetics by the depletion method depends on your technique and how fast your kinetics are (Ch. 8, Ref. 7, 14).

Brash: Yes.

Baquey: There is another technique which can be used to label some proteins and which is used in our laboratory to label antibodies. This method is based on the coupling of radioactive Indium by a coupling agent known as the DTPA (Ch. 1, Ref. 13). As far as I know this coupling agent has the advantage to be fixed mostly on the Fc-fragment of the IgG; so the antibody keeps all of its specific activity against the related antigen.

That's the case for several antibodies which have been labelled in our laboratory.

Beugeling: I think it is good that we are talking about labelling or non-labelling, but we have to fix whether or not we are speaking about single protein solutions, mixtures of proteins, or plasma.

We have to decide what to do, because, if you have a single protein solution, there are many detection methods, and it is not necessary to label, but when you want to measure protein adsorption from mixtures of proteins or plasma, then there are only a few methods which are available. I think that's the point.

Missirlis: Could there be a comparison between the labelling technique here and what the Siena group is doing?

Beugeling: It is not meant for me to answer that question, but in my opinion it is not possible to measure protein adsorption from complex mixtures, such as plasma, with FT-IR if you want to detect one particular protein.

Magnani: Why is it not possible?

Beugeling: Because the mixture is so complex. How do you solve the problem: a certain amount of fibrinogen is adsorbing from plasma to a particular surface and is this amount more or less than the amount which adsorbs to another surface? I don't see how to do it.

Magnani: I have just a little experience in this field, but at Battelle's Laboratory a group of researchers have been working very hard on blood-surface interaction by using FT-IR technique.

Several experiments about protein adsorption from whole blood have been performed there. I can also mention some references on it (Ch. 10, Ref. 4, 5, 7, 8, 9).

Beugeling: It is not a quantitative method, and then I should say that the enzyme-immunoassay (EIA) is much easier; there we have the same problem, because we cannot quantify the amount.

Magnani: Yes, protein infrared spectra cannot be easily distinguished by using peak heights or areas above standard baseline. Therefore the development of characteristic spectral 'fingerprints' for specific proteins or classes of proteins have to be based on the shapes of the more intense bands (Amide I and II) and/or the intensities of the weaker Amide III bands.

Qualitative and quantitative methods for estimating the composition of mixtures of proteins from their IR spectra have been developed and the most useful quantitative methods appear to be based on the matrix algebra.

Anyway, I think FT-IR quantitative analysis can be considered if it is correlated with concentration measurement determined by other quantitative methods. But this, in my opinion, should be a general rule.

Jennissen: Radioactive labelling of protein is also very important for other measurements, specifically for the monitoring of protein desorption. In most cases you can only measure desorption if you radioactively label your protein, because the amount of the desorbed protein is very low. Desorption measurements have been neglected in most protein adsorption studies. Equilibrium desorption studies as well the desorption kinetics are very important for the interpretation of the overall protein adsorption results.

Cazenave: Then you desorb the label and you have to prove again that your label is linked to the protein. You have to have a method and there are methods to make sure that it's the correct molecular weight and it's the correct protein.

Jennissen: Yes, you have to have your controls, of course.

Hoffman: This reminds me of something you said before. Tom Horbett developed a nice technique: page-gel-electrophoresis. He slices it up and cuts the different state levels, and this shows you whether your protein is whole or not.

Baszkin: Chromatography is of course a very good method to separate the free label from the labelled protein, but we believe that a long dialysis is maybe still better, because the small label will always go out to the very big volume of external buffer of water.

Cazenave: Well, you can have proteolysis of your protein. You may have a small fragment bearing the iodinated tyrosine residue which is clipped out of the surface, so you lose radioactivity from the surface, but 99% of the protein is still bound. Hence you have to make sure that the label is linked to your protein.

Missirlis: And how can you be sure of that?

Cazenave: Well, you have to have ways of doing electrophoresis, immuno-blotting, chemical techniques.

Brash: A simple method is, just as Adam said, to dialyse. If you are doing a desorption experiment, you can take the eluate and dialyse it and measure the radioactivity inside and outside the bag.

Cazenave: Yes, but still you have to see in what form it is, so you have to be sure that what is in the bag is, let's say, albumin.

Baszkin: This is a very crucial problem, but this is another question.

Jennissen: In this case an enzyme is an ideal molecule, since it allows you to measure the specific catalytic activity before and after desorption. If the enzyme has remained in the same structural state (i.e. conformation), then the specific catalytic activity should have remained the same.

Cazenave: But again you can desorb the radioactive material, the denatured protein or the active enzyme, so you may have all the enzyme desorbed by the surface, but not enzymatically active.

Jennissen: Yes, you have to perform the proper controls. If you have a radioactive labelled enzyme, it is quite simple. The *specific radioactivity* allows you to measure the amount of protein, and the *specific catalytic activity* yields information on the native state of the enzyme.

Hoffman: If you've radiolabelled a protein, you can run it on a gel-electrophoresis, then slice the gel, radio-count the slices and make sure the radioactivity appears in the appropriate size.

Missirlis: Everybody agrees on this last statement?

Jennissen: Yes, but this gives you no information on the native state or function of your protein.

Hoffman: The question is whether the label is still on the protein, regardless, and whether it's still active or not. That's the only question.

Baszkin: Well, there is also the experiment we use, with the surface tension. If you take such a plot with the native protein, non-labelled protein, and you obtain this curve, and then after labelling, if something happens, you will not obtain the same curve.

Wolf: What happens?

Baszkin: If you take the native protein and you make such a type of plot and you take then the labelled protein, and you may have, let's say, the rupture of some linkages of protein, or, you may have the protein denatured or the protein put into small pieces, or, well, there is still label left, then you will not obtain the same curve.

Because the surface properties of such a protein will not be the same as the surface properties of the unlabelled protein.

Missirlis: Adam, I remember Pr. Wolf asked you whether it would be easy to make the same experiment with both labelled and unlabelled proteins, that you did with ^{14}C, if it's possible to do it with Iodine or Indium and see that you get the same data, so that your data are independent of the labelling technique.

Baszkin: Of course it is possible.

Missirlis: Has it been done?

Baszkin: Not in my lab.

Jennissen: Another important problem concerning the native state of proteins is the problem of proteolytic nicking (limited proteolysis) most often due to the presence of trace amounts of endogenous proteases in the preparation. This results e.g. in the dissociation of the protein into fragments after electrophoresis in the presence of sodium dodecyl sulfate. In physiological solutions it stays together as a holoenzyme and you practically see no differences in the physical parameters although you may have drastic changes in function.

I can give you an example with phosphorylase kinase (Ch. 8, Ref. 26, 27). You can nick this enzyme and it shows practically the same behavior on gel filtration columns and in the ultracentrifuge. However the specific catalytic activity increases 10-fold which indicates something has happened. If you try to separate the subunits by electrophoresis in the presence of SDS they can dissociate into fragments. This demonstrates what large effects the simple proteolytic nicking of a protein can have. You will not be able to detect this phenomenon e.g. by just measuring the specific radioactivity of a labelled protein, because that will remain the same.

I would like to ask a general question. Is there a general agreement on the necessity of cold labelling a protein, if you want to synthesize a protein tracer e.g. by labelling with ^{125}I? As a consequence this would mean that you should dilute your 'hot' labelled tracer protein only with 'cold' labelled protein and not with unlabelled protein. Is there a general agreement on this?

Missirlis: We haven't discussed that.

Brash: What we have discussed are simply ways to validate labelled protein in relation to unlabelled protein, not in relation to cold labelled protein, as you call it. We simply need to have some criteria which will validate the use of the labelled protein as a 'tracer' of the unlabelled protein.

I'm not sure whether your technique will simply say that the labelled radioactive iodinated protein (if we take iodine as an example), behaves the same as the labelled non-radioactive iodinated protein. They are both iodinated proteins.

Jennissen: Why shouldn't the hot and/or cold iodinated species behave the same?

Brash: Oh, I'm not saying it won't. But it's still iodinated, and I think the point in this discussion is that we want a proof that the labelled protein is a tracer of the unlabelled protein.

Jennissen: Yes, but there are a very few methods which will make the labelled protein identical to the native protein (e.g. in vivo labelling by incorporation of ^{3}H or ^{14}C labelled aminoacids). Every chemical labelling procedure will lead to a different molecule of your labelled species, in comparison to the native molecules.
There is no way around this. Then the only possibility would be the in vivo labelling of a protein with ^{3}H or ^{14}C labelled aminoacids (see above).

Hoffman: You have to compare the activity of a protein labelled with either 'hot' or 'cold' labels (as iodine) to see if the radiolabel affects its activity.

Jennissen: Still you never get the same species of molecules in your solution, if one molecule is hot (labelled) and the other one is cold (unlabelled i.e. not cold-labelled). You cannot compare them.

Hoffman: Philosophically you're quite right. There is nobody who can dispute that.

Jennissen: No, not just philosophically, but practically!

Hoffman: But it may not be practically important for a certain test, such as adsorption.

Jennissen: Yes, but I think there are two different things here. In the one case you get preferential adsorption or desorption of your labelled species versus the unlabelled species and in the case of cold-labelling you have a homogenous population of protein molecules, which of course may have an altered function, but which will behave homogeneously in an adsorption experiment.
I have nothing against using even a denatured protein for adsorption or having a protein with an altered function for adsorption, as long as this alteration is defined and the protein preparation is homogenous.

Hoffman: Suppose it does behave homogeneously, but not in the way that would be instructive as to the surface you are testing. Then you've answered the question you don't want the answer to.

Cazenave: To reconcile you, I think you need two controls. You need to dilute your labelled protein with cold-labelled protein and with non-labelled protein. So you will see if the two situations behave in the same way.

But, of course, it depends on what function you are measuring. Let's take for example thrombin. You can label thrombin, and we know, many people know, that thrombin binds to surfaces by its receptor, when it's intact enzymatically or when it's labelled on the active side with DFP, for example. The binding site is not the enzymatic, the serine protein active side, but if you change just one residue you can have a very different function.

So, it depends what is the aim of your experiment. Of course doing experiments is manipulating, and you may never have a proper control. But you cannot do otherwise.

Missirlis: Any more comments on this? Can we conclude on something?

Brash: Nobody mentioned ellipsometry yet, I don't believe.

Beugeling: I thought we are going to talk later in the programme about identification methods, otherwise we can talk now about reflectometry, ellipsometry and so on, but I thought that was planned for later on.

I think we have already heard FTIR and ellipsometry, but I think a much easier technique is reflectometry (Ch. 17, Fig. 1), which gives you nearly the same information and which you can carry out much easier. A student who was working at Akzo Corporate research in Arnhem used a reflectometer for measuring protein adsorption, even from plasma. That was in fact the first step of the EIA; she measured the increase of weight, or the increase of thickness of the protein layer on the surface.

Brash: The only reason I mentioned ellipsometry was because we had a discussion about in situ measurement techniques and ellipsometry certainly is one. It has its limitations, but it is in situ.

Beugeling: Reflectometry, too. It's a similar technique, but, as I have already said, it's much easier to do.

Hoffman: Let me just make you understand something. You are talking about taking a mixture of proteins, such as plasma, and identifying one specific protein coming out, without any label on it, just by the thickness?

Beugeling: Without any label, just by using specific antibodies.

Hoffman: I didn't hear that.

Beugeling: Of course, you have to rinse the surface. There is another possibility: if you use a flow cell, you can displace your plasma or plasma solution by an antibody solution. This is not in situ, but this is close to, and it may be very easy to do. Therefore I think it is a very promising technique and it will be used in many laboratories within a few years. I'm sure.

Jennissen: And quantitatively?

Beugeling: Not quite quantitatively, but better than the EIA, I believe.

Brash: Quantitation of ellipsometry in terms of mass per unit area on the surface requires that some kind of theory is used to interpret the data. Because the fundamental data you get from ellipsometry, and I would assume also from reflectometry, are physical properties of the layer: the thickness and the refractive index, which you get from a change in polarization of the light before and after reflection.

And so, mass of protein on the surface is not something that you get as primary data from ellipsometry and reflectometry. And I believe that there is some dispute regarding the interpretation of the data in terms of the mass.

Missirlis: So, Alan, could you make a concluding or near-concluding statement on the subjects we talked so far?

Hoffman: I really don't think I am the one to make the concluding remarks. They probably should be made by those who are more involved in different isotope labelling.

I can make a few comments: that the commonest, most popular method of labelling is ^{125}I. It seems to work well.

There are ways to assure that you have purified it by getting rid of potential artifacts, such as labelled iodine ions or even iodine.

There are techniques such as ^{14}C and tritium, that I know much less about; that seems to have also certain advantages in some tests.

If I were to start doing research involving protein labelling for studies of protein adsorption on surfaces, I would use ^{125}iodine, with established protocols that many people have used over the years.

Now, the disadvantage of iodine (and I have to tell you that some of my students don't want to use it) is that it's a very potent radioisotope. You can get it into your thyroid and it's very hard to handle. In some laboratories people stay away from it. But that opens up another discussion.

Missirlis: Any comments on that by John and Adam?

Brash: As a personal summary statement on iodine labelling, and in the spirit of making recommendations to beginners, I should say this is a good technique. It's a sensitive technique, and it's very versatile in the sense that you can study single proteins and you can study one protein or two proteins (^{125}I and ^{131}I) in a very complex mixture.

If you are going to use it you have to pay attention to several things. One is that there are three or four different ways to introduce iodine into the protein, such as ICl, which we have not really talked about that much. ICl doesn't involve a reducing agent directly, although, unfortunately, you don't get very high specific radioactivity with ICl. In addition there is lactoperoxidase, there is iodogen, there is chloramine-T. I think these are the four principal ones (see, for example, Regoeczi, E., *Iodine Labeled Plasma Proteins*, Vol. 1, CRC Press, Boca Raton Florida, 1984).

A beginner should be aware of the differences that one can get using those different techniques, what the problems are, what the advantages of each technique are. I would pick lactoperoxidase for the simple reason that there is a lot of literature on the condition of proteins that have been labelled by this technique. It's gentle and you can label at extremely high specific radioactivity, because every iodine atom that you introduce into the protein is radioactive.

With ICl you have both unlabelled as well as labelled ICl molecules. So you end up with a protein of relatively low specific activity. The beginner should be aware of this. He should also be aware of the fact that any investigator using iodine labelling should check a few things regarding the condition of the protein after he has done the labelling.

He should label the protein as close as possible to the time when the experiment will be done. He should check the percentage of unbound iodide which is in his protein preparation as he begins to do the experiment. And possibly at the end of the experiment, as well.

As has been stressed by J.P. Cazenave, it may be the iodine label is no longer attached to the protein at the end of the experiment. Perhaps the protein becomes fragmented.

I guess this is rather a long summary statement, but these are the kinds of things I think one should recommend to a beginner in this field.

Lemm: May I add simply one? You have two iodine isotopes and you can even do competitive studies.

Brash: You can do it two proteins at one time in complex mixtures, if you want. ^{131}I, of course, has a much shorter half-life than ^{125}I and therefore introduces difficulties; your counting equipment has to be more sophisticated to handle two isotopes at one time, but it certainly is possible, although maybe more costly.

Baszkin: When I have described to you our method, you asked me many questions whether the labelled and unlabelled protein behave in the same manner.

I would recommend exactly this same procedure to all people working with ^{125}I, because the problem is the same. And I remember the paper published by the Dutch group (Ch. 2, Ref. 6), in which, for the first time, it was shown that there might be a preferential adsorption of iodine-labelled protein to a given surface. Of course in this case you obtain meaningless results. This is my first comment.

The second comment will be to point out once again the importance of the loosely bound fraction of protein. This loosely bound protein exchanges with the adsorption medium differently than proteins which are close to the surface and which can be altered by their contact with the surface. Also the loosely bound proteins may be polluted by other constituents of the adsorption milieu. It is therefore important to know how much of protein adsorbs in a loosely bound configuration.

Cazenave: The definition of loosely bound?

Brash: What is the definition of adsorption?

Baszkin: What is indicated by an arrow (Ch. 2, Fig. 3) is the collagen which is irreversibly adsorbed to PE. This is the amount of collagen left after the sample was rinsed with a buffer.

When the same experiment is done in the presence of a second protein in the solution, let's say albumin, this quantity is not the same, because a part of this collagen is chased by the second protein, which competes with collagen to occupy the available space or site on the surface: So, 'irreversibly' has different meaning.

Jennissen: The problem here is that the equilibrium parameters were not determined. In each case you should obtain a true or an apparent (Ch. 8, Ref. 15) equilibrium concentration of free protein which is determined by the binding constant.

The only difference in your experiments is that you have conditions of high affinity and low affinity binding. You apparently call high affinity binding 'irreversible' because you could not measure the equilibrium concentration of desorbed protein. If you measure this you will find an equilibrium concentration. It is only very low because the binding affinity is high.

Wolf: You are speaking only in terms of equilibrium, but I think in protein adsorption-desorption processes also the velocity constants are very important.

Jennissen: Well, the affinity is also reflected in the rate constants.

Could I make that comment to the new theme 'suggestions for beginners'? Even on the danger of repeating myself I think one should also suggest to the beginner that, if he is going to label his protein, he should dilute his labelled protein only in a 'cold'-labelled protein. If he does this he will steer clear of a large number of errors.

134

Baquey: I intend to make two comments: First, a technical one dealing with the double labelling. I mean labelling of two proteins in order to carry out competitive adsorption studies.

I would advise beginners that they have to take into account the Compton contribution of the gamma-emission of the [131]Iodine to the photoelectric peak of the [125]Iodine and it is not too easy to do, if the radioactivity of each of the isotopes are not in a good ratio. You have to give advantage to the one to the [125]Iodine in order to lower the contribution, the ratio contribution of the Compton effect of the other isotope. So that is the first advice.

The other comment I would like to do deals with the control of the quality of the protein. All of the discussion has dealt with the in vitro experiments, but if you do in vivo experiments you have another way to control the quality of the protein; this way is to look at the half-life.

If you do the counting of your label in the whole blood, you must have first a curve, which is a b-exponential, which correspond to one, which way correspond to, a great chance to correspond to only one biological specy, and the half-life corresponds to the second part, to second exponential must be that which is well-known for the protein you are working on. I mean that fibrinogen about 3 hours. If you have much less, you have probably a protein which is fragile and which is not in a good state, in as good state as the native fibrinogen.

Cazenave: I have four small comments. Going back to the methods: the lactoperoxidase technique, I agree with you, is a very gentle one, but you have to make sure that you remove the lactoperoxidase and that you don't measure adsorption of lactoperoxidase on your surface. Iodogen is nice because the catalyst is bound on the surface of the tubes are easy to separate from the reactants.

I recommend a book by E. Regoeczi on iodination of proteins in a CRC Critical Review published in 1984, which is quite interesting. I think it contains a lot of fruit for thinking for the beginner.

To have a native protein that is difficult to do in humans, but can be done for special purposes, if you label a protein and you inject it, the damaged protein will be removed from the circulation by several mechanisms, so you end up with a native-like protein in the circulation.

If you measure adsorption of protein from a complex mixture, or plasma, you're sure that you have the remaining protein in the good conformation. So, it's one way to screen out the damage you have done during labelling. It's just leaving the body to do the job.

Brash: In my list of recommendations, I think I omitted to mention the test which Adam does and which I do and many other people do. This is not a surface tension test. It is rather to measure the adsorption as a function of the ratio of labelled/unlabelled protein at the same total solution concentration. If you get a horizontal, straight line in your data of adsorption versus fraction of labelled protein in the solution, the validity of using that labelled protein to measure adsorption is established. It's very easy to do and beginners probably should do it.

Hoffman: This is a classic isotope effect test and everybody should do it when they're starting in this process to make sure they've done it right the first time.

14. Problems of protein adsorption studies – Lyman hypothesis

Brash: I think that somebody who understands what the Lyman hypothesis really is, should make a statement of it, perhaps the person who wrote this on the program. I think that's a good basis to begin the discussion.

Lemm: As far as I understand the Lyman hypothesis: it says that the preference of a surface for albumin is identical for good thromboresistance and in opposite that preference for other proteins, for example fibrinogen or gammaglobulin indicates low thromboresistant properties. Our goal has been to develop screening methods to test the thrombogenicity of artificial materials and though working intensively on this field we did not find a correlation between this hypothesis and the thrombogenicity, neither did we succeed in finding thromboresistant materials. Instead, however we found materials with a high preference for fibrinogen with excellent thromboresistant properties.

Brash: That's also my understanding of the Lyman hypothesis but I felt that it should be stated so that everybody is on the same wavelength. I think that on balance the evidence which has appeared in the last few years and in fact some of the previous presentations support that global hypothesis.

Albumin is good, fibrinogen is bad. Basically that's the hypothesis. And it seems to me that some of the things that we heard yesterday and other things that are in the literature continue to support that point of view. Many people continue to try to develop surfaces which will selectively adsorb albumin from blood, and which would continue to adsorb it after turning it over.

There seems to be at least some consensus among the community that that's a valid approach and a desirable thing to do. It seems to be fairly well accepted that albumin is a passivating protein and if one could develop a surface that would exclusively bind it and continue to bind it, this would be a good surface. Whether it would be the ultimate antithrombogenic surface I think we don't yet know but it's a good working hypothesis.

The other side of the coin, namely that fibrinogen is bad, seems to be in some question. Somebody referred to the work by Lindon and Salzman which appeared in 'Blood' about three years ago (J.N. Lindon et al., *Blood* 68, 355,

Y.F. Missirlis and W. Lemm (eds), Modern Aspects of Protein Adsorption on Biomaterials, 137-148.
© 1991 *Kluwer Academic Publishers. Printed in the Netherlands.*

1986). The conformation or at least the orientation of fibrinogen on the surface and not just simply the quantity of fibrinogen that you find on the surface in contact with blood, related to platelet binding to the fibrinogen.

It seems to me the albumin part of the hypothesis is alive and well, but the fibrinogen part may be somewhat in question at the present time.

Hoffman: I think that we learn more and more about the fact that when you put albumin on a surface it may change in conformation with time and that change will depend upon the surface and the conditions of adsorption, and then the conditions after which it's exposed to.

If you preadsorb albumin on a surface and then expose it to a mild condition such as just PBS, and let it sit for a period of time, we know that its elutibility, its ability to be desorbed, for example, with SDS solutions reduces as time increases, and that means the albumin is changing in conformation.

So if you want to preadsorb albumin as a passivating surface in a fashion that it might change its conformation with time, the question remains, when is albumin a passivating protein and can it change into a non-passivating, or less passivating protein with time? Or more passivating? In fact, we don't know.

Bob Eberhart has been trying to put long alkyl chains on surfaces and his hypothesis is that that kind of a surface will readily exchange native conformation albumin with the surrounding blood albumin such that the surfaces is continuously regenerating a fresh native layer of albumin, which is the passivating protein.

That's the Lyman hypothesis. So, the bottom line of what I'm saying is, I think the Lyman hypothesis has to do with albumin in a native state on a surface and if it does denature or unfold on that surface. Then, I question whether that's still valid. I don't know the answer to that question. We're trying to study it ourselves.

Wolf: I think if you want to speak about this effect then at first we should speak or we should include also to discuss the question why is albumin reducing cell adhesiveness or, better to say, what kind of cells are influenced by an albumin adsorption layer.

We've done, a few years ago, studies to understand this adhesion or cell adhesion reducing effect measuring the influence of albumin adsorption layers to red blood cell adhesiveness and to understand this in terms of the interaction of biophysical forces.

So, we did measure the detachment force for red blood cells and we did measure also the surface charges and also the surface charges of protein-coated layers. It was found that albumin adsorbed onto glass or polymers changes the surface potential only at about 20 mV and makes the surface more positively. That means in principle that the cells should stick more at such surfaces.

After that, we performed studies together with Gingell in London to measure the distance of adhering red cells at the substrate. Normally between glass and the red blood cell we found a distance at about 140 A and after albumin adsorption, this distance was increased at about 40 A. So, the 'albumin cell adhesion reducing effect', in terms of red cells adhesion, means to increase the distance between the interacting surfaces. And it can be physically explained.

But if you're speaking about thrombocyte adhesion onto albumin and fibrinogen then you have to include the question or answer the question 'are there fully different adhesion mechanisms? 'And of course it seems to be clear and it was discussed that platelets stick to fibrinogen by means of a lock-key mechanism. May be there is some part which is also explained by physical forces and so the question arises how we can compare these two types of molecules to answer the questions in terms of cell adhesion.

So, I think we should include in this discussion also the knowledge that we have at this time about cellular adhesiveness on such surfaces.

Baszkin: I would like to remind you that Lyman's (Ch. 2, Ref. 7) hypothesis is based on the irrefutable experimental evidence. This means that there were grafts implanted into the animal body, in contact with blood, and then the protein were desorbed from these grafts and analysed through electrophoresis. It was observed that when the blood did not clot the grafts contained on their surface mainly albumin.

So, I only partly agree with you, because it is very difficult to compare the experiments in vivo and in vitro and I believe that in in vivo experiments the albumin coated surfaces which did not clot could not be compared with any other in vitro experiments. That's why I personally think that as long as another evidence from the in vivo experiments has not be produced to contradict this hypothesis it is very difficult to say that this hypothesis is valid or completely invalid.

Jennissen: At this point I would like to go into the mechanism of cell binding to substituted surfaces and give an example. As you remember the lattice site binding function leads to sigmoidal curves.

There has been a group which has immobilized galactose on a surface, and then adsorbed cells (Ch. 8, Ref. 28). They were hepatic cells which had a galactose binding protein on their surface. They find exactly the same binding behaviour with cells as we see with protein (Ch. 8, Ref. 24). They obtained a sigmoidal curve of cell binding, which indicates multivalent binding. That means that the surface concentration of the galactose or in your case the surface concentration of the protein is decisive for the adsorption of cells.

I think this is a basic mechanism valid for proteins and cells and corresponds to what we have called multivalence. Mechanistically this means the number of protein molecules immobilized on the surface which are capable of simultaneously interacting with the cells is decisive for the binding of cells to the protein-coated surface.

Hoffman: Let's assume that albumin is indeed a passivating protein. Why? There have been a number of arguments put out relating to the simplest in the early days that it was the most negatively charged protein, and therefore, there was a kind of electrostatic repulsion of most everything else which is negatively charged.

There is another hypothesis that relates to its ability to bind fatty acids. I'm not sure exactly how that might work, but at least that's something unique about albumin. Another thing is it's practically devoid of carbohydrate content. And to what extent the carbohydrates participate in binding of other cells, for example, is a question that you just alluded to with your data. And there may be other hypothesis. This is just the question that one would raise rather than answer.

Cazenave: I think for me it's very interesting to understand what you say, because for a biologist, we think in terms of receptors on cells, so if a cell wants to bind to something, it has to have a receptor for the ligand on the surface.

And passivating for what? Let's say for platelets. There is no receptor site for albumin on platelets but there are receptor sites for fibrinogen, for vWF, for thrombin and for a number of molecules.

We've seen, for example, that if you go to a surface with antithrombin III, it's a very passivating protein. Platelets will not interact with the surface. There is no receptor on the platelet.

You could say, but why can you measure some binding of platelets to a surface which is coated with albumin, although this binding is very low? Well, the answer might be that during the process of interaction with the surface, platelets may contact the surface and not stick to it, but be activated and release some of their own fibrinogen from granules. So you are not sure, although you do not add fibrinogen from outside, if you deal with platelets, that they may release traces of fibrinogen on the surface and stick to it.

Beugeling: I fully agree with J.P. Cazenave but, we must remember that there are many other proteins or substances which cannot interact with platelets because the platelets have no receptor sites for them. Yesterday I already mentioned HDL. If you precoat a surface with HDL, platelets will not stick to the surface. Simply in my opinion because the platelets have no receptor sites for HDL.

You can also precoat a surface with cells; we did that with endothelial cells.

Perhaps you say: yes, endothelial cells, that's something else because they release a substance which inhibit platelet adhesion. But, we did the same with smooth muscle cells and found nearly the same results (Ch. 4, Ref. 14; Ch. 19, Fig. 2). Unless you activate your platelets with calcium inophore, for instance. Then you see a difference between the platelet adhesions; there are practically no platelets on the endothelial cells and you see a small number of platelets on smooth muscle cells (Ch. 4, Ref. 14; Ch. 19, Fig. 3). So there is a difference between these cells, probably because endothelial cells are active cells with respect to platelet adhesion and aggregation. But if you only look at the passivation, you can take a substance which does not interact with receptors on the platelet membrane.

Brash: Although platelets are extremely important in the business of designing blood compatible surfaces, I think we're tending to forget here the other aspect of blood compatibility which is of course coagulation. And therefore in the context of Allan's challenge to us to discuss the reasons why albumin might be desirable or an effective protein to have on a surface, we should also be thinking about how does this protein inhibit or even prevent entirely coagulation from occurring on that surface. In general it's axiomatic, I think, that any surface will initiate and promote coagulation to some extent. So that's something else we should discuss in the context of the Lyman hypothesis.

Hoffman: Thank you, John. That was exactly what I was going to say, after Beugeling's comment: Could you, for example, take an HDL and put it on the surface, and would it prevent fibrin formation on that surface? Why is albumin so special? Why do people use albumin to block adsorption sites in chromatographic columns in immuno-diagnostic 96 well-tighten platelets. This is a question that I think relates to its passivation (towards thrombosis).

Baszkin: I think that the idea of precoating polymers with albumin is not an adequate example, because if you precoat in vitro and then you put in vivo this precoated polymer is not the same as the polymer which selectively adsorbs albumin out of blood. The precoated albumin and the albumin which is adsorbs from blood onto the polymer are not the 'same' proteins. They don't have the same configuration and they are not in the same state.

Brash: Perhaps I'll come back to my original point, namely that the Lyman hypothesis needs to be carefully defined. One definition might be, as you say, that thromboresistant implanted materials will select albumin from the blood and they will continue to turn albumin over, as Allan said, with respect to Bob Eberhart's work. That's really the 'pure' Lyman hypothesis: the surface will select albumin from the blood, and will turn albumin over continuously so that there will be a fresh layer of native unaltered protein on the surface.

But people have misinterpreted that and have gone to approaches where you take the surface and precoat it with albumin, perhaps not even the same species albumin and certainly not endogenous albumin. And that's another approach, it's another hypothesis. And as you say, it may well not be valid at all.

It does appear to have some validity based on J.P. Cazenave's experiments on the plasmapheresis filters, where you precoat with, I don't remember exactly what the type of albumin was that you used. . .

Cazenave: The albumin you inject in humans: purified albumin from fractionation of plasma.

Brash: So there is at least some validity to that approach in some applications. But we have to make the distinction, I think, between endogenous albumin picked up from the individual's blood and precoated albumin.

It is unreasonable to expect that you can precoat the surface such that this layer will remain on the surface indefinitely in the same 'passive' form. It seems more likely that this layer will gradually be removed as 'denatured' by normal metabolic degradative processes.

Jennissen: How long is such a surface – which supposedly adsorbs albumin selectively from blood – really the same. When the initially adsorbed albumin molecules leave the surface other molecules are non-specifically adsorbed via their hydrophobic patches or pockets. The surface would therefore change with time. It appears you would need a very highly selective surface – maybe as selective as an antibody binding to albumin – to really get a system like this to work.

Hoffman: You could also crosslink the albumin on the surface to try and hold it there. I asked J.P. yesterday, exactly how long it was after he put the albumin on there that he put the filter in the blood. And I think he said it was instantaneous.

Missirlis: Alan this crosslinking won't affect irreversibly the functional state of albumin?

Hoffman: Treating a protein with glutaraldehyde, for example, always leaves some questions about residue, such as aldehyde groups and such. Depending on whether you kill off all those with ethynolamine or glycine or whatever, you still raise more questions than you answer by doing that. But you can form a layer of albumin, and crosslink it on the surface, and that would be much more permanent on that surface relative to resisting exchange with other proteins.

Baszkin: I would like to say that probably one of the main questions for which we do not have a clear experimental answer is why blood clots in contact with the surface. In this respect, a clear answer should be given. I do not try to defend Lyman's hypothesis but again I would like to remind you that this hypothesis is based on what was done in the in vivo experiments on PU grafts of different chemical and surface compositions. Certain PUs were shown to selectively adsorb albumin from blood and others to selectively adsorb fibrinogen.

Cazenave: As you are challenging the whole conception of coagulation I think people understand that there are at least two ways to start coagulation. One, which is the ex-vivo tube way on a surface, is to first adsorb contact protein and initiate coagulation through the contact activation phase.

This might not be the most important in vivo in humans unless you insert artificial vessels, heart valves or catheters. In vivo it seems that the most important mechanism is when you damage a vessel you release tissue factor and you start coagulation through what is called the extrinsic pathway of coagulation.

So, I think, going back to what Alan Hoffman was saying a surface is passivating to what? To the coagulation system or to cells? If it's to the coagulation system it might be that if you put a layer of albumin on the surface you prevent the adsorption of contact phase activation protein, which will trigger coagulation, so you generate less thrombin and you don't have formation of a clot.

You also prevent the adsorption of platelets, and we know that the platelet surface is important to assemble some of the vitamin K-dependent proteins, which will speed up coagulation because it's a surface reaction. So I think it's a rather complex situation.

Jennissen: It's even more complex if you remember that the complement system is there also, which may also be activated. So, actually, there are many factors present which can influence your system. So the question is: against what are you passivating?

Brash: Focusing back again on the question of why does an albumin layer passivate towards coagulation, you've thrown out the idea that it simply prevents the access of other proteins to the surface, especially if it's in the right conformation and perhaps in the right packing density which covers the surface very effectively. But one can ask a question stemming from that comment: why are other types of protein layers also not passivating towards coagulation.

Cazenave: Gelatin does the job quite well.

Brash: I believe transferrin is also somewhat passivating at least towards platelets. Coagulation I'm not sure.

Beugeling: Everything has already been said. If we want to have a thromboresistant surface or a surface which is biocompatible we have to prevent both platelet adhesion and surface activation of blood coagulation. And we have also to prevent the activation of the complement system via the classical and the alternative pathways; and all these processes interact with each other. So we have to do a lot of things to make a biocompatible surface. That's the problem.

Hoffman: To bring up again sole question which J.P. just brought out: Is a passivating surface one which just simply does not adsorb coagulation proteins or complement proteins that also does not adhere platelets? If this is the case then albumin is just one example of a family of potential polymers, synthetic as well as natural, which could resist that adsorption and adhesion. Among these polymers is PEO and dextran, the two that we are working on. So, this question is still being answered. There is not an answer to this at this time.

Brash: Just a comment with respect to the dextran. I think that it has been shown

144

that dextran is a strong activator of the complement system, because of mechanisms involving hydroxyl groups of which there are many, via the alternative pathway. So I'm not so sure that this is going to be a bland surface at least in that respect.

Wolf: I would say this strategy cannot be the only one to passivate the surface, maybe by albumin. If you want to prevent coagulation or blood cell adhesion then we have to know the special mechanisms which are acting at different molecules.

There are different strategies to achieve it and in case to prevent clotting or complement activation then you should avoid that the special molecules which are adsorbed at the surface and can be surface activated. You can do it may be by a preadsorbed molecules or you can prevent it also by a distinct water layer or maybe also by distinct steric structure of the surface.

It's the same game with cell adhesion. If you want to prevent cells to adhere at a surface the first step will be a biophysical or physical step that means electrostatic repulsion. So we can prevent every cell maybe by electrostatic repulsion to approach to the surface. And then the second step is how this special type of cell is really adhering. We have to make differences between red blood cells, platelets and white cells concerning its adhesion mechanism.

And so we can create a distinct electrical structure at the surface and we can also construct a special receptor structure at the surface without or with a coated protein. Also the fluidity of cell membrane is an important point and one cannot speak only about the number of receptors at the cell surface. If there is a change of the fluidity of cell membrane there could be also a change of the adhesiveness of the cell. So, what we should do is to recognise these biological principles and then to try to model such bionic structures at our material surfaces.

Baquey: I would like to comment about two examples related to ex vivo experiments; we observed a lot of fibrinogen adsorbed or retained by the surfaces which were less thrombogenic than the control.

The first example is dealing with heparinized catheters which have been developed by Rhone-Poulenc and studied in Rosy Eloy's laboratory. These catheters are less thrombogenic than non-heparinized catheters made of the same material. That means that the time during which blood is able to flow through them is longer for these catheters than for the others. But if you look at the fibrinogen, you have a lot of fibrinogen which is depositing on the wall.

And parallely you can see and Rosy Eloy did this work, there is no FpA which is generated, (very low level compared to the control catheters). So in this case there is a lot of fibrinogen without resulting in immediate thrombosis.

The other example is the example given about the so called heparin-like materials. In this case you have less fibrinogen which is adsorbed on the control tube than on the treated tubes. And the treated tubes have kinetics of platelet adhesion which is less than on the control tube. So in this case, too, fibrinogen doesn't seem to be the condition of thrombogenic phenomena.

Beugeling: I have a comment on your last example because the heparin-like material you use is negatively charged. Perhaps less cell adhesion occurs because the surface is negatively charged.

In the case of heparinized surfaces there is always the discussion:is there a leakage of heparin or is there no leakage. Some people are saying there is no leakage and nevertheless the surface is antithrombogenic. But I'm not quite sure that there was no leakage.

Baquey: No, I would like to make some addition to my previous talk. My question is not to compare the two types of surfaces. I want to give two examples in which there is objectively accumulation of fibrinogen on the surface, but in one case there is leakage of heparin that has been well demonstrated with well known kinetics and in the other case we have a surface which is able to adsorb specifically AT III and to catalyse the normal inhibition reaction of the thrombin by the AT III.

Brash: I can comment additionally on those heparin-like surfaces. You mentioned a moment ago that they adsorbed specifically antithrombin III. But in fact we've done some collaborative work with Marcel and Jacqueline Jozefowicz which shows that although antithrombin III is there, there is a lot more fibrinogen adsorbed than any other protein when you expose those materials to plasma (Ch. 5, Ref. 1).

So there is AT III but there is perhaps 100 times more fibrinogen on the surface, so I'm not really sure how you can correlate the behaviour of those materials only to antithrombin III being on the surface.

Baquey: Yes, I quite agree. I was only saying that on the control which has not specific affinity for the AT III there was less fibrinogen which was deposited than on the material which was specifically designed to adsorb AT III and which featured ex-vivo a lower thrombogenic potential. And the results I am talking about have been obtained in vivo.

Brash: This could be documented as another counter example to the fibrinogen part of the Lyman hypothesis, I believe.

Baszkin: There is also one thing that should be added. The behaviour of heparinized or heparin-like surfaces, in contact with blood is different than the behaviour of polymers which do not contain heparin, or which do not contain heparin-like molecules or groups. The adsorption of proteins out of blood should be different on what we call inert polymers, and polymers which are active.

Consequently, different classes of polymers in contact with blood should be divided onto those which have no heparin or heparin-like molecules on their surfaces, and these, which can selectively adsorb different proteins out of the medium.

146

Baquey: Yes, but the question was: is fibrinogen able to induce thrombosis? And I gave two examples, in which it is clear that fibrinogen was adsorbed in vivo; it is clear again, that this fibrinogen is not, (at least for the first example) fragmented in order to give rise to fibrin and in this example also the material which accumulates more fibrinogen than the control is less thrombogenic in vivo than the control. So, it's totally in contradiction with the Lyman hypothesis.

Lemm: So, may I ask a question for the future? Is the Lyman hypothesis really a useful tool in terms of designing new and non-thrombogenic surfaces?

Brash: I think we should perhaps generalise the Lyman hypothesis. It has been stated in terms of albumin and fibrinogen. It seems to me it needs to be generalised. I believe strongly that specific proteins on surfaces do specific things, subsequently, when that surface is exposed to a biological system.

So, I would like to propose that the hypothesis be revised to say that in protein adsorption, the identity of the proteins in the layer are important. Indeed they may control entirely the subsequent interactions of that material with the biological environment.

We don't have a range of knowledge which tells us what each of the proteins that might be there will do. But, certainly, the composition and character of the protein layer is outcome-determining for the biocompatibility of that material.

Clearly we are not in agreement with regard to what different proteins do. Except perhaps for albumin. It's my feeling that it's still valid to say that albumin is desirable on the surface. I'm not sure I would go much further than that.

Hoffman: I quite agree. I think the hypothesis is still valid. The way that I would generalize it is that it is valid to search for a surface which is repellent to proteins and cells, with the hypothesis that such a non-fouling surface would be non-thrombogenic.

I have a five year grant from the NIH to study non-fouling surfaces and hopefully to improve those which are already known as PEO. So, we continue in this area very heavily.

Lemm: Why, do just the polyurethanes adsorb preferably the albumin? And why is their blood compatibility better than that of other materials? Can this effect be explained by protein adsorption tests?

Beugeling: You said polyurethane (PU) adsorbs so much albumin, but that's only relatively. I think that the amount of albumin is very small compared to other proteins, so you can only say this PU adsorbs a little bit more albumin than other PUs do. But whether that's the reason that they are blood compatible, I don't know.

I personally, think that the polyether groups on the surface, which are rather hydrophilic, are oriented in aqueous media in such a way that you have a rather hydrophilic surface, and that it is the hydrophilicity of this surface, which makes it more blood compatible. I don't know whether or not others agree with this.

Brash: I would also challenge the premise that these PUs, in general, are strong adsorbers of albumin. I am not familiar with any convincing evidence to support that point of view.

I would say that, in general, materials that you place in contact with plasma or blood, have a tendency to adsorb very little albumin.

When you place the polymer in contact with plasma or blood, the amount of albumin that you are going to find on the surface, compared to other proteins, is going to be very small. And I think that's correct for PUs as well as for most other surfaces. So, I am not sure that the premise here is correct.

I would also challenge the idea that PUs as a class, are all that much more blood compatible than many other materials that one can think of.

It seems to me that the reason why these materials are popular and are used a lot is simply from a mechanical property stand point. They have tremendous versatility. You can make them with a wide range of mechanical properties.

But, in the compatibility aspect, I'm not so sure that they are all that outstanding.

Beugeling: From some in vitro experiments, we know that platelet adhesion is less on Pellethane compared to silicone rubber (SR) or low density polyethylene (LDPE). (Fig. 1). But you can even make it better by making it more hydrophilic. And that we did by grafting polyethylene oxide (PEO) with a molecular weight of 180, 000 on the surface and then you see less platelet adhesion than on Pellethane itself (Ch. 4, Ref. 18).

And the reason for this is, in my opinion, simply because this material adsorbs less protein. You don't see any albumin on the grafted surface.

Brash: I can mention something else, if you want to stimulate discussion on PUs, and that is the Japanese hypothesis about PU and block copolymers (T. Okano et al., *J. Biomed. Mater. Res.,* 15, 393, 1981). This says that there is a mosaic of hard and soft segment domain structure on the surface of the PU due to microphase separation.

The mosaic surface has relatively hydrophilic and relatively hydrophobic patches, which have the same order of magnitude dimensions as the hydrophobic and hydrophilic patches which are present on a cell surface.

Therefore, when a cell comes to the surface of this microphase separated material, it finds a distribution of hydrophobic and hydrophilic patches which match closely to those on the cell itself. Therefore, the cell may sit down on the surface but that's all that happens.

It's just like a billiard ball coming and sitting on the surface. It just sits. It has no tendency to spread or undergo any kind of physical deformation.

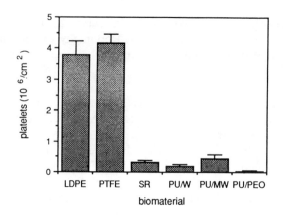

Fig. 1. Platelet deposition to various biomaterials from perfusates containing plasma ([citrate] = 13 mM). Shortly before perfusion 1 μM Ca^{2+}-ionophore A 23187 was added to the perfusates. Perfusions were carried out for 5 min at shear rates of 300 s^{-1}. All values are mean ($\pm$ SD) of four determinations.

So, that's the Japanese hypothesis and I suppose we could discuss that if you want to discuss PUs in terms of why people use them and why people think they are compatible, if they are.

15. Polymer surface properties

Lemm: Well, I would like to discuss surface properties measured by all kinds of tests and its correlation to protein adsorption.

Jennissen: It appears that the main problem is that most of these surfaces are not characterized.

Take for example the agarose gel: we have an X-ray crystallographic study of the agarose (Ch. 8, Ref. 2). We know the exact spacing of the helices, we know the exact spacing of the agarobiose units and when we immobilise residues we know how far they are apart. We don't know if a residue is up or down. We have made scanning electron microscopic (SEM) pictures of agarose where you can see the porous and the non-porous areas where proteins can be adsorbed (Ch. 8, Ref. 4).

A major problem is that most surfaces are not characterized concerning protein adsorption. So, you can say very little about the mechanism of protein binding on these surfaces.

Hoffman: I think that a lot of work is going into characterizing surfaces, but not necessarily in the molecular level that you're suggesting.

There is a lot of evidence that polymer surfaces of different compositions adsorb different proteins differently. But there is no unifying theory that will allow you to predict if you are able to identify the surface composition on a particular material how it would behave in contact with polymers.

There are some generalities that you know already:that hydrophilic surfaces, surfaces that tend to adsorb water, tend to be fairly slippery toward proteins, slippery meaning proteins come down, have a relatively short residence time in a shear field, and then they leave. This is what that suggests.

Hydrophobic surfaces with a layer of strongly bound, very low entropy type of water, will then encourage protein adsorption and strong adhesion because of the displacement of that bound water which is driving forces I've mentioned earlier.

The words 'hydrophilic' and 'hydrophobic' are used, I think, so often because you have, if you take the contact angle of water on a surface, a really

Y.F. Missirlis and W. Lemm (eds), Modern Aspects of Protein Adsorption on Biomaterials, 149-161.

© 1991 *Kluwer Academic Publishers. Printed in the Netherlands.*

hydrophilic surface that allows the water droplet to spread completely, in fact, to adsorb into the material. Yet, there are many surfaces with different contact angles of water, ranging from low angles, all the way up to 90°. These surfaces are of varying degrees of hydrophilicity as you get towards the lower angle on.

So, you have to be careful in using that.

I want to mention also the other factors on a surface, which are important and sometimes overlooked, are the molecular mobilities of the molecules that make up the polymer itself. You can have a polymer which is effectively hydrophobic, such as silicon rubber which has a very flexible chain. You can have a polymer which is effectively hydrophilic and has a relatively stiff chain, such as a cellulose membrane for dialysis. Nevertheless, it has much water in it and therefore the chains are relatively mobile, but the backbone itself is not very flexible.

In addition to molecular mobilities you have the concept of movement, a physical movement of the surface itself. If it's a rubber and has the ability to bend or is being pumped as a heart valve or a diaphragm that movement can influence the surface adsorption of protein significantly.

You can go beyond the question of mobility on a molecular or microscopic sense to the question of topography or roughness. You have surfaces continually imagined in our minds and on blackboards as flat lines with water molecules or proteins sitting on them and they are not flat. Even a gas-discharged treated surface is not flat on a molecular level. We don't know what they look like at the molecular-level scale.

So, this is another consideration, particularly hydrophobic surfaces with small imperfections that can trap air bubbles which can lead to denatured proteins and platelet aggregates and so on.

Then you can go beyond this to the question of domains we were talking about with PU: the idea that the surface composition, whatever it is, is not uniform across the surface. You have compositional changes across the surface whether you put them there or whether in fact they got there, because your fingerprint was on it. They might be impurities, they might be oxidation, they might be domains of PU and the like.

So, you have, in fact, the idea that your roughness, your mobilities, crystalline regions, amorphous regions, all of these can vary across the surface. Thus these idealized pictures and imagined ideas about the surface being in uniform composition (that is flat and either hydrophilic or hydrophobic), are very simplistic.

We have to be very careful about dealing with them, but on the other hand, we can be over concerned about this situation and never do anything. In the end, I recommend that one concentrates on ESCA as a means for analysis of surface composition.

Hopefully we are developing techniques that can be used to analyze these surfaces which are in contact with water. There are low vacuum, high vacuum, low pressure systems, usually in the dehydrated form and that's not the natural condition. So, we are trying to work with frozen water.

SIMS is another very useful technique. Contact angles are also very useful and simple, but you have to be aware that when you put a drop on a surface, as you put it down and then pull away from it, you have actually forced the water or the drop of liquid to advance over the surface and then to retract. So, you have to be careful of that, and this comes out from a recent talk by George Whiteside at the Biomaterials Conference.

But ESCA, SIMS and contact angles are useful; FTIR I find a little tricky. We don't use it very much, but it's certainly useful in the hands of many people, and some have, like radiolabelling techniques, their personal preferences.

Baszkin: I didn't understand this comment of Whitesides. Can you repeat it please?

Hoffman: It had to do with the technique, of putting a drop on a surface in order to measure contact angle.

You take a drop in a teflon syringe, where you don't contaminate anything. You have a long tube with a tiny point on the end and you're going to come down to the surface and put the drop there.

How do you get your needle or syringe away without forcing the liquid down on the surface as you're leaving, and thus you're getting an advancing and receiving angle?

Beugeling: I think the key point is perhaps to keep the pipette tip in contact with the drop, when you do the measurement, and not to separate them.

Hoffman: You don't need to take the syringe away. You actually measure the drop contact angle when the syringe is still there. Then you pull back and the liquid sucks on.

Baszkin: All surface chemists are doing this way.

Brash: Many people do contact angles by the so-called classical sessile drop technique which involves the separation of the syringe or pipette from the drop and that's exactly what he said one should not do. Because then you get a mixed advancing and receding angle.

Beugeling: I fully agree with Prof. Hoffman; we also do ESCA and we measured hydrophilicity or hydrophobicity with contact angle measurements. But we used the captive bubble method, in order to avoid some problems and because we think that is a more real situation. In blood or tissue the polymer is always in contact with an aqueous medium so I think you can better use the captive bubble method for the contact angle measurements.

And one remark about ESCA: we also identify groups on the surfaces especially the treated Teflon surfaces. The only thing you cannot say is how they are oriented on the surface. This orientation may be very important.

Hoffman: Just to follow up. I quite agree with you. Underwater measurements of contact angles of captive bubbles or even hydrophobic liquids like octane are much more relevant than air-surface measurements which sometimes are easier to do, and that is why so many people do them.

Brash: I think those techniques that people have mentioned so far like ESCA, SIMS, and contact angle are the techniques that we have at the moment. Those techniques are available to most people – maybe not ESCA, but certainly contact angle and maybe FTIR also.

It seems to me, though, that we need some other techniques to track certain problems and to get information that current techniques don't give us.

And one type of information that we don't get with current techniques is a mapping of the surface, an area distribution of chemical composition.

For example to go back to the case of the segmented PU with microphase domain structures. These microphase domains exist in the bulk of the sample but we really don't know what the domain structure is right on the surface, because the domain mapping is done by EM, and usually that involves transmission right through the sample, so it's a bulk measurement and we really don't know what the surface distribution of domains is. People have hypothesised regarding that but have not really been able to measure it.

So techniques are required not only for segmented PU but for any material if you're interested to know the chemical mapping of the surface. It's very difficult, if not impossible to get it, with high resolution, using currently available methods.

Baquey: I think there could be a solution at a microscale level which is not of high resolution. We've been able to propose mapping of the surface through the technique of autoradiography. If the surface is polyanionic a mapping of such a surface can be obtained.

That is the case for dialysis membrane such as polyacrylonitrile based membranes developed by Rhone-Poulenc; you may get a first mapping of the surface with ^{45}Ca at a microscale level, using the same technique as used for fibrinogen (Fig. 1).

But you can imagine to do the same thing, using Transmission Electronic Microscopy, and developing the technique of micro-auto-radiography, but at a higher resolution level.

And the best for that is to introduce during the synthesis, if you think of PUs, ^{14}C precursors in order to assess the distribution of the microdomains on the surface; you will get a print of labelled segments on the autoradiographic document at the resolution which is given by an electron microscope.

Lemm: Since a few years, a new method is available. It's called the tunnel electron scanning microscopy. What is your opinion on this? Is it a tool to identify a surface?

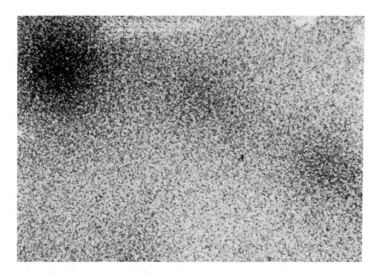

Fig. 1. Microautoradiography of a dialysis membrane (HOSPAL-AN 69) previously incubated with a buffered solution of ^{125}I-labelled fibrinogen (size of the explored area: 50 x 80 μm^2).

Hoffman: This is something I was just going to mention. We have a scanning tunneling microscope at the University of Washington and we are at present trying to develop techniques for imaging individual proteins on surfaces.

I think these are games right now. They are not serious yet in terms of real surfaces, the problem being that you need a conducting surface in order to observe a particular image. So the tendency would be then to use a conducting surface onto which you would deposit a protein in solution and you would look through that solution on the surface and image the proteins as you scanned across. That means that you have to look at surfaces like carbon. Eventually we can begin to look at protein solutions.

Charles Baquey reminded me of another technique called colloidal gold imaging. I've seen some interesting studies recently by Albrecht and Cooper, where they used colloidal gold onto which they had adsorbed fibrinogen to mark the fibrinogen binding sites on platelets.

Hoffman: ... a function of time and to see as a platelet spreads on a foreign surface, where these particular GP23A glycoprotein binding sites are with time as this platelet interaction with the surface, and try and correlate this with surface properties. It's a very interesting technique and again it might be used. I haven't seen much of it being used for mapping where fibrinogen is going down on a foreign surface as opposed to this platelet surface.

Brash: That works after the protein has gone down on the surface, but if you want to characterise the material itself it doesn't work, it can't work.

Hoffman: The fibrinogen is already adsorbed on the gold's particle.

Brash: For example if I want to map a PU or any other material before it has been implanted to know what the distribution of chemical composition is on the surface I have to find another way to do it.

Jennissen: In many cases the problem is that you first get your material and then afterwards have to analyse it. In our case, for example, we have our alkyl residues radioactively labelled. And in making the surface ourselves we can easily determine the surface concentrations of the immobilized residues.

So there are techniques if you're capable of making the surface yourself of determining: 1) the spacing between certain molecules, 2) the function for example, if it's hydrophobic or hydrophilic or if it's charged. And maybe this techniques should be employed more often also with the surfaces that you are using.

Baquey: But in this case I suppose you get a statistical distribution, but not a real mapping as required by John.

Jennissen: Yes, we obtain a statistical distribution of the immobilized residues.

Beugeling: I quite agree that you can determine the amount of groups on the surface. We isolated factor VIII by special Sepharoses which were modified. The simplest way is putting aminoalkyl chains on the surface. Then you can simply titrate your free amino groups to see how much of them are bound. So, often, there are easier techniques to see what is present on the surface.

Baszkin: I wish to draw your attention to the technique which we have in our laboratory which enables the in situ quantification of the number of functional groups in contact with an aqueous phase by adsorbing radioactive cations or anions. An exact quantity of the effectively present functional sites turned towards the aqueous phase can thus be measured.

Different techniques of surface chemistry different spectroscopies and different microscopies provide a more or less exact information of what is at the surface of polymers. The problem is that these different techniques are not available in one single laboratory. I think that Prof. Hoffman's laboratory is one of these laboratories which are best equipped. But there is no one single laboratory in which all these techniques are present. That's probably one of the reasons why a full characterization of polymer surfaces in contact with an aqueous phase is very difficult.

Hoffman: May I just emphasize that our ESCA facility, we have ESCA and SIMS and now a scanning tunneling microscopy) is supported by the U.S. government with in fact a very nice donation from the Shell chemical company. But, it's a national service center. It's open for collaborative research. It's open for your own studies for you to send samples and or to come and do your own samples there. And the service is relatively inexpensive. I think, in fact, it's free for scientific researchers. Buddy Ratner is the director of it and if you're interested in more information I can provide that or I can ask Buddy to send it to you. So, it's not just our facility. It's supported by the U. S. government for the use of researchers and I believe around the world.

Brash: I can confirm that they don't worry about being American to use this. As a Canadian I have taken advantage of it.

Koutsoukos: I would like also to point another factor which should not be overlooked when we are looking at surfaces and this is surface charge which probably can be looked rather easily by electrophoretic techniques or also by potentiometric titrations to find proton titrations. And with respect to Dr. Hoffman's observation for Scanning Tunelling Microscopy I would like to ask if conducting polymers such as polyaniline or polypyrol have been looked at for investigating conformation on the surface of such conducting polymers.

Hoffman: I didn't hear the conducting polymers that you mentioned. What were they?

Koutsoukos: I mentioned two widely used. One is polyaniline and the other is polypyrol.

Hoffman: Even though I wanted to hear their names, the answer is I don't know. This is an interesting possibility to look at conducting polymers. There are some available: pyrolized polymers and so on. But, in the end, I don't know any commonly used biomaterial that's normally conducting. And that's the problem.

Jennissen: What hard information on the later protein binding of the surface can you really gain by the ESCA method? How relevant is this information e.g. for the number of contact sites between the protein and the surface?

Hoffman: Your question has to do with molecular binding of an individual protein molecule to a surface and what you can learn from ESCA about that? I don't think ESCA can get down to that level of knowledge. ESCA can give you information sometimes about the tendency of proteins to adsorb in uniform or more or less monolayer versus multilayer islands by doing angular dependant ESCA studies. But in the sense of how many binding sites per protein, I haven't seen anybody get that kind of information out of ESCA.

Brash: I think it's precisely because the spatial resolution is not very great on ESCA.

Hoffman: 50 μm.

Brash: Yes, the spot size is huge compared to any protein molecule. Could I raise another aspect of the chemistry of surfaces, which I'm not sure that we've really addressed, i. e. functional groups? I think that ESCA can give some information on that. As well as just atomic composition it can give you information regarding the bonding states of the atoms and therefore the chemical functionalities.

One aspect of chemical functionalities is how the group is attached to the surface, and again Allan Hoffman and others have done work on this in the past, using spacer arms, for example, to make the functional groups more available to whatever species in the biological medium they are intended to interact with.

It seems to me that's something we should be discussing at this point in this session. There is, for example, discussion as to what is the optimum spacer arm length, if you want to have a specific group such as a sulphonate or an amine group on the surface, which will interact with a given protein or, perhaps, with a cell receptor.

Beugeling: In our lab Riethorst did a rather long study to see what kind of material he had to use to bind as much as factor VIII from human blood plasma as possible (Ch. 4, Ref. 15). For instance if you take the aminoalkyl residues. Then there was an optimum in the recovery of factor VIII for n (number of C atoms in the alkyl chain) is about 3 or 4. If you make longer chains then you bind more proteins because there is also an hydrophobic interaction between the long hydrophobic chain and protein. His idea was: the aminoalkyl Sepharoses are not good enough for binding factor VIII because you must also elute factor VIII. He made long hydrophilic spacers, which were more effective in isolating factor VIII from blood plasma. With the same amino group on the end.

Jennissen: Ch. 8, Fig. 1 (Ch. 8, Ref. 6) shows the dependence of the adsorption of a large molecule, phosphorylase-kinase (m = 1.300.000) as a function of the surface concentration of alkyl amines. This is the butyl residue, the ethyl residue and the methyl residue. It also shows that even immobilized methyl groups can adsorb a protein if you obtain the right surface concentration.

So the answer to the question of how long a spacer arm has to be to be able to interact with a protein is that right down to the methyl group they will interact if you have a sufficiently high surface concentration.

Wolf: Can I ask you one question? If you use then another type of protein molecule for adsorption will you get then a fully different picture? You know, in the case of blood, that we have about 150 proteins and so if one makes one group for one molecule, how the selectivity can be guaranteed?

Jennissen: The selectivity of binding is different all proteins having a different surface topography. For different proteins the critical surface concentration of immobilized alkyl residues, (butyl-, ethyl- and methyl-) necessary for adsorption is different (Ch. 8, Ref. 6). It depends on the number of interactions ('multivalence') between the protein and the adsorbent surface (alkyl residues). If you compare phosphorylase with phosphorylase-kinase the sigmoidal curves and the critical surface concentration of residues are different, since the minimum number of alkyl residues necessary for the adsorption of these proteins is different. Phosphorylase kinase is adsorbed at much lower surface concentrations of alkyl residues than phosphorylase.

Hoffman: In terms of biologic activity of immobilized drug molecules such as heparin or the like, I believe that Kim and Fejen have shown that the optimum spacer arm is about 6 to 8 carbon atoms in an alkyl group. This is in relevance also to biomaterials. That's for immobilization of biologically active molecules such as antithrombogenic agents.

Brash: The question to be addressed is the availability of chemical groups on a surface to interact. I think there have been many studies of chemical functionalities on surfaces, e. g. sulphonate groups, sulphonamide groups, carboxylic acid groups, amine groups etc. And I think very often people have been misled into saying that certain groups are good, and others are bad. But they really haven't constructed the material in such a fashion that you can really say anything about the interaction of the functional groups. Because unless the group is available, unless it's hanging in the wind, so to speak, with respect to the blood, it's not going to interact. Clearly the same argument applies to immobilization of molecules like heparin, prostaglandins and the like.

Missirlis: Yesterday we had the question whether we should use commercially available or from our own laboratories proteins. Does this apply also to the polymeric surfaces? And these IUPAC materials are they characterised according to this discussion here well, or they need more characterisation? Could somebody comment on that?

Wolf: I could say they are well characterised. The companies which were the producers of these materials have sent also a protocol about the data which they have got. So, ESCA was used, SIMS was used, colloid chemical methods and so on. All points discussed here I think were given into the protocol for the primary reference materials.

Lemm: Prof. Baszkin suggested a very nice method to identify the functionality of the surface. Is everyone familiar with this method? I know this method and I find it's very nice with the radioions.

158

Baszkin: As you know, since 1970, we are developing different in situ methods. I have presented to you, during this workshop, the in situ method to measure adsorption-desorption of protein at interfaces.

Earlier we have developed an in situ method to quantify the number of functional groups at the polymer surfaces. This method gives you an immediate answer for certain types of problems.

Here is an example: You take a PE on which you graft acrylic acid. Polymer chemists weight the sample after grafting and they know the increase in weight. They say, we have grafted, let us say, 10% of acrylic. However if you bring such a surface in contact with the solution of calcium ions containing $^{45}Ca^{++}$, you discover that there is no adsorption of these ions. All grafted COOH groups are inside of your PE. This means at the surface you have nothing.

Lemm: And by ESCA what did you find?

Baszkin: By ESCA you can find acrylic acid.

Beugeling: In this case, there is a much easier test. You can measure whether or not the contact angle has been changed.

Baszkin: Yes, but contact angle measurements are not enough sensitive as a test. You have to work very carefully with contact angles to have an irrefutable answer to such a delicate problem. With $^{45}Ca^{++}$ adsorption you have an immediate answer.

Baquey: Maybe I could give a precision. I think that the beta energy of ^{45}Ca is about 240 KeV. And under these conditions the length, the mean range which can be crossed by the particles could be about 50 μm in a material with the density of 2. So it could be larger in PE. So, I think the volume explored has a contribution. Under these conditions, whichever is the measurement, you take into account a layer with a 50 to 100 μm depth.

Baszkin: You subtract the backround and you can know the quantity which adsorbs.

The best way will be to look at Ch. 2, Ref. 3.

Baquey: No, that's the same...

Baszkin: But of course this method has also its limitations. We use ^{45}Ca and thiocynate labelled ^{14}C for such type experiments, but in all these cases, you have also limitations, because the surface must be able to adsorb Ca^{++} or you must have functional sites which adsorb Ca^{++} or another functional sites which do adsorb monovalent anions.

You cannot check all types of polymers. You can check only the polymers which were grafted with, let's say, ionic monomers.

Jennissen: I would like to come back to the so-called 'spacer arms'. The points I wanted to make a minute ago were just to show you that the protein can see a methyl group. Whether it will interact with it depends on the surface concentration. Recognizing this principle will however allow you to control such unwanted interactions.

Baszkin: I have no answer on your question. But before coming to this particular problem I think that protein adsorption is not only governed by acid-base interactions or by electrostatic interactions, but also, and to a very large extend, by what we call long range forces, or dispersion forces.

So, you can imagine, that even if the functional groups are buried in the polymer bulk, if the spacer is not exposed to the surface, even if it is in the bulk, changes the long range forces of the material towards the aqueous solution are changed. So it plays a very big role.

Lemm: Any more comments?

Baquey: Yes, once again, as it was said before by J. Brash, I think the mean density of the specific sites it important, but has not been further analysed. You need a nice distribution, like in a crystal lattice, in order to get the wished behaviour of the protein, or of the biomolecule which is to be adsorbed. I think that's the point.

Jennissen: There is a statistical distribution of immobilized alkyl residues and the adsorption of protein molecules is also statistical, as in any chemical reaction. On this basis you can make certain conclusions on the binding mechanism on the surface.

Baquey: When you 'do polymer chemistry' that's different when living organisms are 'doing polymer chemistry'. It's totally different. When any polymer chemist does polymer chemistry he may get statistical modification.

I mean for a surface you may describe a surface as a grey surface. That is because grey is a mixture of white and black. For example, check boards are made of black and white squares; these ones may be large or small squares, but 50% of the surface is always black or white. But, according to the size of the squares, and according to the square 'seen', the surface is not at all the same for the molecule or the cell which is looking at the surface.

Jennissen: Yes, but at the moment we're far away from having a surface with an orientation of all immobilized groups in only one direction. You would expect something like this on a crystal surface. The statistical surface is the one we have now. Can you suggest a better one? One that's practical today?

Baquey: I've nothing to suggest, but I'm aware of the difficulty.

160

Jennissen: I just want to remind you that the basis for the Langmuir isotherm is also statistical.

Baquey: But biological behaviour is not statistical. That's the problem.

Baszkin: I would like once again to point out that the presence of specific functional groups at the surface of a polymer or in general of a material do not necessary enhance or block protein adsorption. If you change the bulk of the material, you still influence the properties of the material by changing the contribution of long-range forces operating at the material-aqueous solution interface.

Wolf: Can you characterise what you mean if you're saying long range forces?

Baszkin: It needs of course a long discussion and a long presentation, but very simply, if you have a solid material in contact with a liquid, there are interfacial forces operating at the solid-liquid interface.

When you analyse these interfacial forces in the terms of the solid-liquid work of adhesion, you can roughly divide them into two categories: those which are due to the Van der Waals forces (long-range forces) which are always present and those called polar forces or acid-base forces, which can be present or maybe not present at the interface.

And if you modify a material, by incorporating into it another one, you modify the contribution of the long range forces.

The long range forces would have an important effect on adsorption of proteins. An introduction of a spacer or its modification would directly influence upon the long range forces and so on adsorption of proteins.

Wolf: May I make a comment? Dispersion forces are short range forces, because the electrostatic forces may be acting until, say, 1000 Å, in maximum.

Baszkin: The Van der Waals forces are considered to be long range forces (Ch. 2, Ref. 8).

Wolf: No, I don't know that, I think it's not...

Missirlis: That's a good point to come into an agreement or disagreement.

Jennissen: Electrostatic forces are proportional to the reciprocal of r^2. The Van der Waals forces are proportional to the reciprocal of r^6. So you have a dramatic decrease in the attractive force of Van der Waals interactions . Could you explain your point again on the long range forces?

Hoffman: It's quite often $1/r^3$ for long range forces. It's in between the two of the Coulombic and the Van der Waals. But I really wanted to make another comment and that is, that all of this is relevant in the sense that the protein or the biomolecule approaches the surface which, again, is not flat.

It is a molecular level that we are approaching. There are holes in it, and there are holes between the molecules of the polymer. So you are going to sense from a distance, as we sense the gravity of the earth, longer range attractive forces.

Also, if you're coming in the aqueous solution, you're going to potentially allow rearrangements to occur at that interface as the protein approaches to minimize the interfacial free energy.

This process on a molecular level is very complicated and I can't imagine it very easily. It's simply described as the polar groups coming out and the non-polar parts of the polymer going as far away from the water as they can.

So, this argument, about short range and long range forces in a sense still imagines the surfaces as being a nice flat surface. I think at the molecular level it becomes much more complicated.

16. In vitro, in vivo or ex vivo studies

Lemm: If there are no more contributions on the topic: 'surface characterization', let us tackle another subject, the correlation between the studies in vitro and in vivo/ex vivo. Most of us perform in vitro studies, except Prof. Baquey. You are the only one who studied protein adsorption ex vivo, which is much more difficult.

The results of in vitro experiments can be well reproduced. Is there a correlation between these in vitro results and yours obtained from in vivo or ex vivo experiments?

Baquey: Sure. I think there are a few points which have to be taken into account at first.

When in vitro experiments are done, apart from the first experiments reported by J.P. Cazenave, using non-anticoagulated blood which are obviously short term experiments, all other experiments carried out in vitro are made with blood or plasma or protein solution which are Ca^{++} depleted or which are anticoagulated.

So, in vivo phenomena which are Ca-dependent or which could be inhibited by heparin are by-passed. And a lot of phenomena cannot happen in vitro as they do in vivo. So, any comparison is difficult. Thus, I think that in vitro experiments can be used for specific purpose of surface chemistry or for comparing several materials, for a given property, but it's difficult to use them to predict an in vivo behaviour.

Beugeling: Yesterday we already talked about the Cazenave flow system (Ch.4, Fig. 1). In that system you can use a suspension of washed platelets which contains Ca^{++} and Mg^{++}. In that case you have your Ca^{++} back. I agree with you, then you don't have plasma but you can put the platelets back in plasma and, for instance, activate these platelets, by adding calcium ionophore.

So, that's just what we are trying to do, trying to correlate the in vitro experiments with in vivo experiments. That's what everyone wants to do. Then we can predict results from in vitro experiments.

Y.F. Missirlis and W. Lemm (eds), Modern Aspects of Protein Adsorption on Biomaterials, 163-168.
© 1991 *Kluwer Academic Publishers. Printed in the Netherlands.*

164

Baquey: We do also the same. And we do a lot of experiments, like you do, and like Jean-Pierre does, because the method was learned from Jean-Pierre. But in our opinion we use this experiment to screen among several materials those which are supposed to be brought to the ex vivo experiment, because ex vivo experiment is very expensive and very hard to carry out.

Wolf: Only a short comment. If you want now to speak about the comparison between in vivo and in vitro experiments then we should at first fix about what we want really to speak. What type of parameters we want to compare?

That means maybe looking only at platelet adhesion or looking only at protein adsorption or comparison of, say, leukocyte adhesion or thrombus formation or activation of clotting system.

Because we cannot make such an overall discussion. We should go at one point what we want to compare really. Otherwise it makes not much sense to do this.

Lemm: Protein adsorption of course.

Hoffman: It seems to me from the general feeling of the people that I know in this field that a whole battery of tests in vitro is needed in order to characterize a material and also to attempt to predict its in vivo performance and correlate with it.

There are so many factors involved here that it's practically impossible to reach a general conclusion and say: fibrinogen adsorption is the in vitro test we want to try to correlate with an in vivo performance.

There are questions of the in vivo animal species, the type of test, where your implant or explant (that is where your ex vivo shunt is going), and what you're measuring. There are so many different questions you can raise that its hazardous to make a general statements. So, be very careful.

I would say: do what you can. If you have a series of tests that you're good at and you have an animal model that makes sense to the end application, then I think it's good to try and correlate.

There are very few correlations which I know of. We can imagine correlations; we can force them. We have a small diameter vascular graft project we worked on for the past ten years. We've shown that certain treatments, such as fluorocarbon discharge treatment produces a vascular graft, a small Dacron blood vessel, which when put in an ex vivo shunt remains open for over a week . If we don't treat it, it's occluded with blood within hours or certainly after a day or so.

So, we've seen an ex vivo shunt performance which is very interesting. We've also seen that on those treated surfaces, albumin not only is adsorbed but it's retained much better than on the untreated surface.

There is a correlation. What does it mean? This is the opening of a whole new Pandora's box, and sometimes you can find a correlation. But it sometimes doesn't mean anything, but we do our best.

Missirlis: Alan you just mentioned that you know of a few good correlation studies in vitro-in vivo. Would you be so kind to put the references in the manuscript so that will be included?

Hoffman: The reason I know about them is that there are so few decent animal models. Cooper and Mike Sefton have some data that seem to correlate. Alan Callow in collaboration with Salzman and Merrill seem to have some useful data. These are people doing animal modelling that may have some correlations in the literature. I can't tell you exactly which references now.

Cazenave: I would like to add a little bit more to these difficulties. We've tried to develop an animal model with small diameter vascular grafts and there you run into difficulties. We have tried to mount an ex vivo model which would correlate with our in vitro perfusion system, using grafts of 4 mm diameter or less. And then you run into trouble, I would say.

First you have to select animal species. Well, the best animal species I know is man, but it's rather difficult to do long term studies, although we're developing with hollow-fibers some ex vivo one shot studies, which are short term.

If you want a longer term you need an animal. A few people have access, as you have in Seattle, to baboons. But they are expensive. They are good animal models, because most of the biological tools (antibodies against human proteins) we have cross-react with the baboon. And don't cross-react with other animal species.

So, we decided to use rats. Because rats are small, they are not expensive animals and you can develop a shunt model which you put in the carotid artery and in the jugular vein. We can have living conscious animals, bearing a 30 cm long tubing of small diameter, which we can maintain for hours.

Then you have a problem as to how you record thrombus formation, which I agree is different from protein adsorption or platelet deposition. In order to do such studies, you need, for example, to isolate proteins, which we are doing. We're trying to have a complete rat system.

We isolate proteins, clotting factors and platelets and make antibodies. So it's a choice; it's difficult. We've rat albumin, we've antibodies, we've rat vWF, we've antibodies, so we can label them, inject then in the animal and record with γ-camera protein deposition, platelet deposition and thrombus time formation which we recorded up to now by measuring the difference in temperature in the shunt. So, you can do a study in one day.

I was talking with Charles and telling him that the chap who has been doing that, called Nicolini, in our lab is a vascular surgeon. You need people with good hands. When he started with PE tubing his observation time of a catheter was about 2 hrs. Six months later, using the same PE tubing, it was 8 hrs.

So, I warn people that may think they have been doing great surface chemistry and great platelet deposition or protein adsorption, you need to have proper controls. And we know with experience how not to kill these animals, to do quickly and not to get thrombin generation at the insertion of the tubing.

This is all quite difficult, but by doing these studies we could show that using different types of biomaterials, albumin adsorption improved the quality of the raw material, fibrinogen adsorption was not good and thrombin or collagen adsorption on the catheter would give thrombosis very quickly.

So, all these are possible, but you need to invest a lot of time if you want to go into an animal model. And I think that for screening purposes some in vitro tests, well, are quite predictive of what could happen later on.

And also, and I'll finish on that point, we know that there are two important things when blood interacts with a catheter. I think clinical and experimental studies have shown that antiplatelet therapy is important. Because antiplatelet drugs are the only ones which will prevent thrombus formation in shunts and hemodialysis circuits. All drug companies have tested their products in this respect and they work.

The second thing is that thrombin is very important. Recent data in Harker's group have shown that hirudin or other specific antithrombin inhibitors prevent thrombosis. So I think that platelets are key and the effect of thrombin on the system and on platelets is also very important. Thus, we have to be very careful when we design our experiments ex vivo.

Hoffman: I'm wondering. John, do you remember? The NIH put out a book I mentioned earlier in the meeting that contains recommended in vitro tests. Do they also have recommended in vivo and ex vivo testing? I think it's there, too. I will put it in the reference.

Brash: I'm sure that mention is made in the NIH publication of in vivo tests that were available. This is a relatively old document, published in 1984, I believe, but it may be a good place to start.

I would like to narrow the discussion. I appreciate all the comments about animal models and difficulties of working in vivo, but maybe one question that we should try to answer here is whether there are correlations in vivo versus in vitro of protein adsorption itself.

If you're doing an in vitro experiment with plasma or some other fluid and you find a certain distribution of protein adsorption on the surface, you would like to know if that distribution is correlated with the same distribution or a different distribution of adsorption in an in vivo experiment. And if the distribution is different, how is it different, and is there any explanation?

Allan, I think your references to correlations in Cooper's work are really more in terms of precoating materials with different proteins and then performing an in vivo experiment to determine the thrombogenicity, or lack or it, as a result of the preadsorbed protein layer. But with respect to actual adsorption of proteins from the blood in vivo, other than fibrinogen, I don't think he is doing that.

Hoffman: There has been very little in protein adsorption measurements in vivo that I know. I think Jerry Shultz had a dog model a long time ago, where he radio labelled with chromium. He had a counter over the shunt and was actually counting the shunt as a function of time. That was an ex vivo shunt, of course. I thought Cooper was doing this with his segmented shunt as well. But I could be wrong.

Baquey: I may add some comments about the Cooper's method. In fact, I don't know other method used in vivo or ex vivo to study the adsorption of proteins or more generally the behaviour of proteins at the interface.

Most of them expose the surface of interest to blood for a given period of time and this period may be short, as does Rosy Eloy in Lyon for catheters, and then the surface is collected and investigated. But I think that the in situ examination could be better.

And Cooper is not a real in situ examination. He uses segmented ex vivo shunt and he collects one fragment, then another fragment in a sequential procedure.

My concern about such a model is that each collection step is separated by a rinsing procedure: so, I'm not sure that the events which are coming after the first rinsing procedure are the same as the events which would have happened if the rinsing procedure had not been carried out.

Brash: This is a personal comment in terms of correlating in vitro with in vivo protein adsorption. I would just like to mention again what we're doing, which is to take dialysers, after clinical use, and elute the proteins from the surfaces of the membranes. One can make a direct comparison between what you see under those conditions, which is as 'in vivo' as you're likely to get, and exactly the same experiment with the plasma from the same individual in vitro.

And that's what we are in the process of doing right now. We have published a paper on the in vivo experiments (Ch. 5, Ref. 16) which were the first ones that in vivo we did, and we're still in the process of doing the in vitro comparison. But this is something that can be done.

One can think of all the therapeutic in vivo procedures that are done in patients involving blood-material contact, such as hemodialysis, heart lung by pass and the like and do a similar experiment.

The main difference between in vivo and in vitro in this experiment is that in vivo we see a lot of non specific degradation of proteins which are deposited on the dialyser surface. We see much less degradation in vitro.

Whether these observations have any great significance in terms of blood compatibility and thrombogenesis I don't know at this point. But these observations have not been made in the past, and may well have some significance. The main point is that one can do this direct in vivo-in vitro comparison by taking advantage of clinical procedures which are being done anyway.

Cazenave: I think one of the difficulties, John, which we all have and which we should be aware of, is that when we collect a segment of material from human

168

or an animal, well, we have to interrupt at some point and create stasis. So you generate thrombin. It's very difficult to know what we should do. In an animal should we give heparin just before collection to prevent any or to reduce coagulation triggering or not? And this might be very difficult.

I would recall – it's not biomaterials and I'll go quickly. When I was at McMaster we were measuring prostacyclin formation by segments of rabbit aorta (Ch. 6, Ref. 14). We realised that there was a huge variation from one animal to another and there was a correlation with the amount of clot in the segment during collection, because we had to kill an animal, clamp the vessel, remove it and rinse. As soon as we gave heparin before collecting the vessel, the basal level of prostacyclin was even from animal to animal.

So I warn people that they have to be careful because you may introduce some artifact which is difficult to avoid.

Wolf: I have one short question to the comment of John Brash. That means, you take dialysers and after that you make a rinsing procedure and make the analysis as shown yesterday and is it clarified or is it sure that what you desorb by your rinsing procedure? Is there not any effect from the rinsing procedure? Does it mean you find some other molecules that they are destroyed during rinsing or is your technique already so clear?

Brash: There is no particular assumption involved in this aspect of the study. We simply say that whatever comes off in the rinse must have been adsorbed on the surface and that's the major point. Whether the rinsing itself can do anything undesirable to the proteins, which would influence the analysis of the experiment, we really don't know at this point. I think one could address the same question to any procedure that we use to study protein adsorption where rinsing is required.

Unfortunately this is the case in most experimental procedures. They do involve rinsing the surface, either to measure bound protein or to release the proteins from the surface.

There are techniques, however, involving antibody probing, which at least don't require you to release the protein from the surface. In our procedure we are not making any assumptions regarding what the rinse itself does. We merely identify the proteins that come off.

17. Kinetics of protein adsorption

Missirlis: The first topic we have, it has the title 'Kinetics of protein adsorption', with three subsections: 'static or dynamic protein adsorption studies', the second is 'experimental strategies to study dynamic protein adsorption' and a third one: 'the influence of flow conditions' – like Re number and others' – on the adsorption behaviour. Who wishes to take the floor? Or to lead the discussion?

Beugeling: I already told you yesterday that of course we want to do dynamic measurements of protein adsorption, but that is not always possible. But you can try to get as close as possible, and in the case of our EIA it is about 15 seconds. This time can be shortened if you use a flow cell. For instance, if you take a small cell and you put in your protein solution, or plasma, and then displace your protein solution or plasma after a certain time by a first antibody solution you can shorten the time very much, I think.

To do the whole EIA in that way is a little bit difficult, I think, because you have to do many things, like washing and adding your labelled second antibody and then your substrate and leuko dye solution and so on. So it is better if you can take another method which can detect the adsorbed protein after the first step of the assay. I think that is possible by doing this with a reflectometer. If you have a small flow cell you put in your protein solution or plasma, displace it immediately by first antibody solution and you are measuring what's going on. Then you can specifically detect one protein on the surface within let's say seconds or perhaps 0.5 sec.

A laser beam is reflected by a reflecting surface (Fig. 1), for instance a chromium plate or a silicon crystal which you can spin-coat with a polymery, so that you have a very thin film. Of course your polymer must be soluble. Because you have a very thin film on your silicon wafer or on your chromium plate, it is still reflecting.

What you are measuring is the reflected amplitude of the perpendicularly polarized light (R_s) and the reflected amplitude of the parallelly polarised light (R_p), and that is performed by the beam splitter and by the two photo-detectors.

Y.F. Missirlis and W. Lemm (eds), Modern Aspects of Protein Adsorption on Biomaterials,
169-199.
© 1991 *Kluwer Academic Publishers. Printed in the Netherlands.*

170

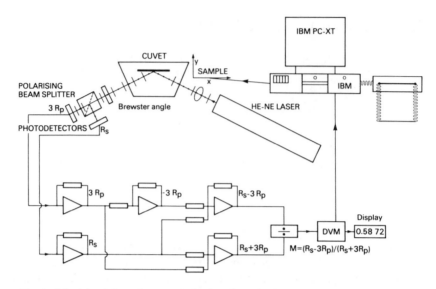

Fig. 1. Principle of the reflectometer developed at AKZO Corporate Research, Arnhem,
The Netherlands.
– R_p is the reflected amplitude of parallelly polarized light.
– R_s is the reflected amplitude of perpendicularly polarized light.
– M is the readout value.
(From Heuvelsland et al., Ch. 4, Ref.16).

At a certain angle, the Brewster angle, R_p is zero, but that changes when
protein adsorption occurs, or when the surface is coated with a thin layer of
polymer. With the electronics you are calculating the read out value M.

The value M is plotted as a function of the layer thickness of the coated film
(Fig 2). Now, suppose that you have 0.1 μm of polymer on the surface. You see
that in this region M is a linear function of the thickness. That means if protein
adsorption occurs you have only to measure the change of M to see how the
thickness of the protein layer is changing.

This is shown in Fig. 3 but it's not measured automatically; you see the
adsorption of bovine serum albumin (BSA), in this case on silicon which was
coated with a thin layer of polystyrene (PS) and this is the oxidized silicon
wafer, so a hydrophilic surface, and you see the increase, in this case of the mass
of adsorbed BSA.

You must take a certain specific density for your protein on the surface; this
does not differ very much from the density of the solution itself because it
contains a lot of water. But assuming a certain density you can calculate the
mass of the deposited protein on the surface. And you see that you get
adsorption curves as we already know.

M. Oldenzeel from our university did her research at Akzo/Arnhem (The
Netherlands) and she used that apparatus and the ultimate goal was of course
to measure protein adsorption from plasma or dilluted plasma.

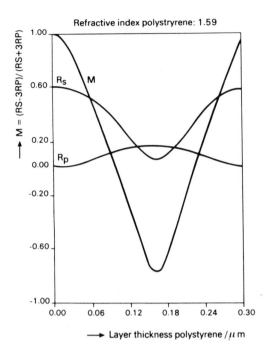

Fig. 2. Relationship between readout value M and the thickness of the polystyrene layer coated on the reflecting surface.
(From Heuvelsland et al., Ch. 4, Ref. 16).

She combined the technique with antibodies. She put the plasma into the cuvette and then replaced it via buffer by a first antibody solution. And that she did for several plasma concentrations, as you can see here (Fig. 4).

As a first attempt to do protein adsorption from plasma I think it is a rather nice result because we all know these curves from other measurements. When you use a flow cell it must be possible to measure protein adsorption after very short contact times.

It is a very easy technique. You can put the apparatus on your desk; it is not sensitive for vibrations.

You can also measure adsorption of proteins from single protein solutions (Fig. 3); you have only to measure the increase in the thickness of the layer. And if you want to measure the increase of the total amount of protein from solution or from plasma you just look at the increase and that has been done here for the whole plasma.

Of course this (Fig. 4, X) is not the sum of these (Fig. 4, △, ○, □) because what you are measuring here is the increase of the thickness of the layer due to the antibodies which are reacting on the surface, and which are specific.

I'm convinced that within a few years many people will use this reflectometer to study protein adsorption. Because it is so simple. It has already been used in

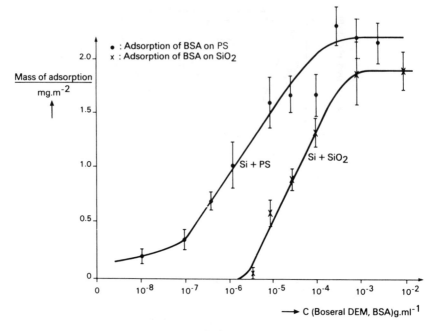

Fig. 3. Adsorption of bovine serum albumin (BSA) from single protein solutions to silicon wafers, either coated with polystyrene (PS) or oxidized (SiO₂). Adsorption time 1 hour.
pH = 8.9 and 22°C.
(From Heuvelsland et al., Ch. 4, Ref. 16).

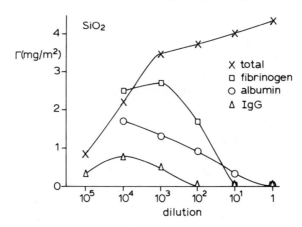

Fig. 4. Adsorption of fibrinogen, albumin, and immunoglobulin G from diluted and undiluted plasma to SiO₂. The total amount of adsorbed
protein is indicated with X.
(From M. Oldenzeel, unpublished results).

1984 by Welin (Ch. 4, Ref. 17); that publication was in an analytical paper. He was the first who measured protein adsorption by reflectometry but at

Akzo/Arnhem they computerized the whole thing. Without computer it costs about $10,000, so nearly everyone can built it.

Hoffman: Tom, I was wondering. Those data were based strictly on the thickness of the adsorbed film, which the implication is then, if you're saturating all of the adsorbed protein with the corresponding antibody, then the Vroman effect is applicable to the immune complex as well as to the individual protein. Which I think is interesting if that's, in fact, what's happening.

Beugeling: The problem is that the height of the adsorption plateau is perhaps not dependant on the amount of adsorbed protein you want to detect but on the packed layer of first antibodies on the surface.

Hoffman: You say the packed layer of first antibody. First you put down the protein, right? Then you chase it with the antibody.

Beugeling: If you use a polyclonal antibody, the density of the antibody molecules on the surface will be larger than in the case of the monoclonal antibody. But if you use a polyclonal antibody, I think the packed layer of the antibody molecules determines the height of the adsorption plateau.

Lemm: But, please, the question this morning should be 'do the results obtained under static conditions differ from those obtained under dynamic conditions'?

Beugeling: The problem is that we want to measure dynamically, because you want to see the first events which happen. You want to know what is happening within a second. Perhaps you can do that with the methods we discussed today.

Brash: I just want to comment that in terms of the schedule, we seem to be addressing the second point which is 'experimental strategies to study dynamic protein adsorption'. Maybe it's O.K. to continue on that and exhaust that topic and then go back to the more global one which is 'what is the influence of flow and fluid dynamics on adsorption if any?

I would suggest that we continue on this topic of methods. I think it is important to delineate methods that can be used to study adsorption in situ and with very high time resolution right from the beginning of the interaction.

Regarding reflectometry, I would like to ask Tom how you convert the thickness data into mass. Is there any difficulty with that, are there any assumptions to go into the calculation of mass per unit area from thickness? The numbers that you showed, around 0.3 to 0.4 $\mu g/cm^2$, were certainly very reasonable. So would you tell us about that?

Beugeling: Yes, the problem is that you have to know the density of the adsorbed layer otherwise you cannot transfer it into mass. But you can do an estimation of the density and do that by adsorbing just one protein from a single protein solution.

Now, from experiments with labelled proteins you know what the plateau value is in mass units. So you can use these data to correlate them with the measurements from the reflectometer. And in that way you can estimate the density of the adsorbed protein layer. Probably every protein has about the same density and this density is nearly the same as the density of protein solution.

Wolf: My question concerns also the technique. We've used internal reflection microscopy also to measure the distance of cells adhering onto surfaces. And then we've used a theoretical model to calculate the distance including the glycocalyx on the cell surfaces. And if I understood your technique correct, then you have to introduce in your calculation the refractive index to get out the thickness or you have to assume one thickness to get...

Beugeling: You're measuring the thickness. That's the only thing you are measuring. Because M is a linear function of the thickness.

Wolf: Yes, but if you use this function as I understood, that's my question, you have to put into the calculation the refractive index.

And that is in my opinion the problem because if you have a changing adsorption layer the first question would be what is the refractive index in such adsorption layer, also monolayer. That should be different from the bulk phase.

And if your adsorption layer is growing that means starting from 0, then maybe you could put in the calculation the same index like in the bulk phase but after that, there should be a continuous change as a function of time and at the end of adsorption, if you have a packed layer as you were speaking, then this refractive index should be different.

Maybe it makes, only, say, 100% of the other value but then your thickness would change also in 100%. and that is the problem involved. What we have done to obtain such values was for instance to model the glycocalyx. We have used different sucrose or other sugar solutions to measure the function of refractive index on the concentration of such sugar solutions.

So something you have to put in your equation to get something out. And so this is similar to the ellipsometry or is it not so?

Beugeling: In principle you can measure more with ellipsometry. People say that they can measure both the thickness of the layer and the mass. But in practice it turns out that you can do nothing more than with the reflectometer.

I agree with you that that's the weakness of the method. You have to assume one thing and that is the density. I'm not saying that it is an ideal method; it's just an attempt to go a little bit further. And it's much simpler than the EIA. Because you have only to do the first step.

Hoffman: I'm not trying to change your apparatus, but what about using an ELISA set up so that you can actually measure the signal change with time. The signal, of course, is developed by an enzyme, so it's a slow process. But is there another way? I guess it comes to Adam's system where you measure radioactivity as a function of time.

Baszkin: I have seen such an apparatus in the Biomaterials laboratory in the Government Industrial Research Institute in Osaka (Japan) (Ch. 2, Ref. 9). They have developed a method of measuring rotational diffusion of proteins adsorbed on polymer surfaces. The method is based on the use of nitrogen laser and on the measurement of time resolved fluorescence anisotropy (in nanoseconds) of a fluorescent probe covalently bound to a protein by a total internal reflexion (TIR) technique. An important information on protein adsorption may be obtained at the very beginning of the process.

Jennissen: Couldn't you just make a calibration curve of your ELISA data versus the data you obtained by reflectometry. This would allow you to make calculations directly.

Beugeling: Perhaps this is possible but we do not yet have a reflectometer. I'm very happy with the results of mrs. Oldenzeel because that encourages me to go further with this method.

Brash: I would like to ask you another question about the method. You mentioned that this works in situ and at very short time. The question is what do you mean by short time? Also when you do the plasma studies using antibodies, then it's no longer in situ because you're introducing first of all your plasma, then you're rinsing or at least displacing the plasma with a buffer or a solution that contains the antibody. So there is a gap between the event itself and the measurement of the event. So, I guess it's a two part question. One part is what do you mean by short time, and the second is the in situ aspect.

Beugeling: We did not yet use a flow cell, but that was my suggestion.

In these measurements a normal small cuvette was used and we know that you can measure with the EIA after about 15 sec; that is the shortest time. But I do hope that you can get closer with a flow cell. Let's say to about 1 sec or 0.5 sec. And of course that's not the beginning of protein adsorption but it is already closer to the beginning.

And if you look for instance at Ch. 4, Fig. 5 you clearly see that from 15 seconds to a few minutes there is a decrease of the amount of adsorbed fibrinogen and also of HDL and probably other proteins, so I think if we go to, let's say 1 sec, we can also see how much is adsorbed at that moment and then you can perhaps extrapolate to shorter times.

Missirlis: The other question. The in situ problem. Can you comment on that?

Beugeling: That's of course a restriction. You must displace your plasma with an antibody solution, or buffer and then antibody solution. I do hope that you can do it in one step. That you can directly displace your plasma or plasma solution by an antibody solution. Perhaps you have to use a higher concentration of antibodies. At the interface there is a reaction between antibodies and adsorbed protein, and there is also protein in the solution.

Hoffman: You know, Tom, I asked a question before about immune complex and the Vroman effect. But it occurs to me that, in fact, what you run the risk of here is the Vroman effect itself where the IgG, regardless of whether it is an antibody, will essentially displace. There is a competition with its binding and it's potential for displacement of the protein.

Beugeling: I fully agree with you that's the risk of it. That IgG is displacing a protein. In the first step of the EIA precautions are taken. You add e.g. gelatin and bovine serum albumin. So, there is a lot of protein in the first antibody solution in order to prevent that the antibody adsorbs at open places on the surface. In this case you can also say that if you have more protein in the solution then there is an opportunity to displace your adsorbed protein.

Missirlis: Are there any other experimental set ups?

Hoffman: I think that the technique allows you to investigate looking down in this first section. I'm not too concerned about whether we're going to mix A, B and C together because they are really interrelated. By varying the shear rate you have the possibility here also of studying the influence of that on protein adhesion, which would be interesting.

This really is dynamic ellipsometry, isn't it? That's basically what you're doing. If you combine it as Jennissen said with ELISA, if you can stop it at any moment quickly, take out your coated disk and put it in a ELISA tray, then you have a very good way to quantitate and correlate this whole system. I think it's a good idea.

Missirlis: In the mean time can one say a few words about ellipsometry? I mean people may not have this background.

Van Damme: I think ellipsometry, the whole system is somewhat more complicated to operate. You have to have an apparatus in which you are changing continuously some polarizing filters, so you have to have a computer who is controlling the whole system. And depending on the speed of the computers the speed of the measurements are changed. So if you have a rapid changing surface the time between the points is longer.

That is the main problem. And there are, in our electrotechnical department, they are now working about a rotating polarizer which gives some more possibilities.

Then you can measure I think every 0.5 sec, but you need a large computer to do all the calculations. So, it's a bit more complicated but you get more data from it. It's not used for polymers and proteins in electrotechnics. They use it a lot for things like oxidation of metals which is quite another system.

Hoffman: I had a question about it. In ellipsometry you normally take, your plate, you wash it and then look at the thickness of the adsorbed protein. How does it work if you look at it in the presence of varying concentrations of protein solutions?

Van Damme: What you usually do, I think is, you don't wash it, you put your plates into a buffer and then you add the protein. And you look at the kinetics of the adsorption.

Hoffman: Basically there is no problem. I'm thinking of your experiment with different shear rates you have different concentrations of proteins near the surface, as a function of shear rate. Then it will still work.

Van Damme: Yes, but I think the best opportunities of these techniques are in measuring single proteins. And when you're going to measure from mixtures you're running into problems because you have different refracting indices of the different proteins, locally. And even in combination with antibodies or with an EIA it's difficult. You need very flat surfaces.

Hoffman: That's a prerequisite. It must be very smooth. How about the index of refraction variation? What is the range?

Van Damme: I don't really know.

Brash: It's my understanding that in ellipsometry, the thickness and the refractive index of the layer are obtained from the change in polarization of the reflected light.

If you're measuring a mixture of proteins you would get some kind of average refractive index of the layer. Then you have to try to interpret that average refractive index in terms of a complex composition. So, unless you have some kind of theory of mixtures of proteins in a layer in terms of refractive index, you would not be able to interpret those data in terms of layer composition.

I'm in the market to buy an ellipsometer, but it's not to study plasma or anything related to blood, it's to study basic protein adsorption mechanisms in single protein systems for which, I think, it's very well adapted. There is well established theory to relate the refractive index and thickness data that you get to the mass of protein on the surface.

So, for me this is a technique which works exquisitely well for single proteins, but not for plasma or anything complex unless a method of interpretation of the data is developed.

I would like to present briefly a kinetic in situ technique, which we have been using recently. We have not published on it yet, but I will give a talk at the ACS Colloid and Surface Science Symposium in Seattle next month.

This is a technique which actually was published by a group in Strasbourg last year in 'Colloids and Surfaces' (J.D. Aptel et al., *Colloids and Surfaces*, 29, 359, 1988). It's a relatively old technique, which surface chemists call 'serum replacement'. Our objective of course is to obtain quantitative (by which I mean mass of protein per unit area), kinetic data in very high time resolution in situ, as the events occur, on systems of interest to the biomaterials research area.

The system is very simple. If you're a chemical engineer the cell is a CSTR, a 'continuous stirred tank reactor', except it's not for a chemical reaction it's for a protein adsorption experiment.

What you have is a vessel with inflow and outflow, which is extremely well stirred, so that the concentration is uniform throughout the volume of the cell and is identical with the concentration of the solution which leaves the cell.

Initially, buffer fills the cell. At a given time, time zero (t_0), you inject the protein solution of interest, and that might be plasma or a single protein solution which contains a protein whose concentration you can measure somehow. The way we do it is by radioiodine labelling.

You inject the protein solution at a given volumetric flow-rate, v_0. V is the volume of the cell and C_p is the concentration of the protein solution in the cell resulting from processes that take place in the cell.

Now, suppose you have particles of the surface of interest, in the cell, being agitated and maintained in suspension. What happens as you inject the protein solution is that you have mixing of the solution into the volume, which reduces the concentration, and you have adsorption of the protein on to the particles, which also decreases the concentration of the protein solution.

You can calculate the quantity of protein adsorbed unambiguously by doing a simple material balance around the flow cell. It's an unsteady state protein balance in the sense that the protein concentration is not constant with time: it's changing.

Fig. 5 shows the balance without particles first of all. $V(dC_p/dt)$ is the rate of accumulation of protein in the fluid volume of the cell, C_p is the protein concentration in the cell. The first term on the right is the rate of protein entering the cell and the second term is the rate of protein leaving the cell.

If there are no particles, then all you're doing is mixing the protein into the volume. You can solve this differential equation very easily and the result is shown.

If you add particles to the system you have the accumulation term, you have the rate of protein entering the cell, the rate of protein leaving the cell and then you have another term due to adsorption shown as $r_{ads}S$. You're removing protein from the fluid volume by having it adsorb onto the surface of the particles.

UNSTEADY STATE PROTEIN BALANCES

Without particles:

$$V \frac{dC_p}{dt} = v_o (C_{po} - C_p)$$

$$C_p = C_{po} [1 - \exp(-v_o t/V)]$$

With particles:

$$V \frac{dC_p}{dt} = v_o (C_{po} - C_p) - r_{ads} S$$

$$VC_p(t) = v_o C_{po} t - v_o \int_0^t C_p(t)dt - S\Gamma(t)$$

$$VA(t) = v_o A_o t - v_o \int_0^t A(t)dt - SA_s(t)$$

$$\Gamma(t) = \frac{A_s(t)}{A_o} C_{po}$$

Fig. 5. Unsteady state protein balances.

And you can start putting in models for this r_{ads} term, Langmuir models, or whatever kinetic model you want to try to test or use.

Alternatively, you can simply integrate this equation as shown. Again we have the accumulation term; the amount of protein that has entered the cell up to a given time t; the amount of protein which has left the cell up to the same time; and the amount which has been adsorbed onto the particles. You can write this equation in terms of radioactivity A, because that's the way that we measure protein concentration. From this you can calculate the surface concentration of protein on the particle surfaces, knowing the radioactivity of the solution and the concentration of the solution in mass of protein per unit volume.

So all you have to do then is solve the mass balance at frequent time intervals and you get a virtually continuous measure of adsorption as a function of time.

Jennissen: What are you adding? Are you adding a constant protein concentration continuously or are you adding a pulse of protein?

Brash: No, you're adding a constant protein concentration. So eventually the concentration in the cell will become equal to the concentration of the solution that is being injected. And when that happens then you can no longer measure anything as you've reached steady state.

Fig. 6 shows the details of the actual experimental set up. You have to have some kind of filter (we use a glass filter), to keep the particles inside the cell when you make the measurement. You can use a syringe pump to inject the protein solution into the cell. At the exit to the cell, which is as close as possible to the cell to minimize 'dead' volume, you place a radioactivity counting system, and then you take the data to a data analysis system.

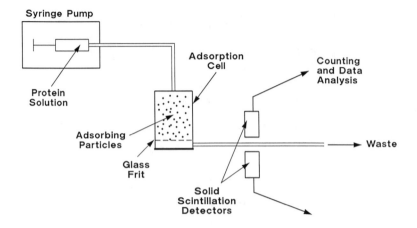

Fig. 6. Kinetic in situ Adsorption experiment.

Fig. 7 shows more details of the data analysis system we use. The protein solution flows from the cell through the tube in which you're measuring the radioactivity. Radioactivity is detected by a scintillation system along with data storage and analysis using a relatively fast computer system.

The measurement really consists of measuring the radioactivity in the exit solution at very short successive time intervals. This is shown in Fig. 8 for a system of glass particles and a pure fibrinogen solution. Without particles you get the upper data set. You can see a solid curve which is the solution to that first equation which I mentioned before. The data follow the theoretical curve quite well.

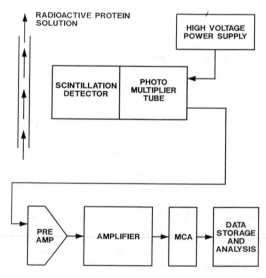

Fig. 7. Radioactivity detection and data analysis system.

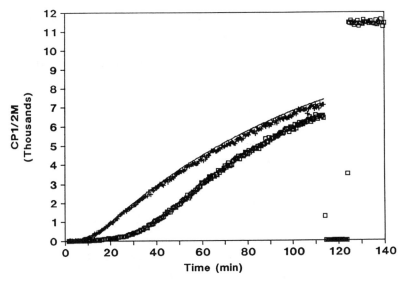

Fig. 8. Radioactivity of cell effluent as a function of time. (C_{f0} = 95 μg/mL in Tris, v_0 = 0.15 mL/min, S = 1030 cm², A_0 = 11451 CP1/2M).

+ + = Activity recorded in the absence of glass beads

□□ = Activity recorded in the presence of glass beads

_____ = Calculated control run

Tubing rinsed with buffer at 115 minutes.

We do a time measurement every 30 seconds but we could do it every 0.1 sec. It's just a question of changing the dwell time on the multichannel analyser which can go down to 10⁻⁶ sec if you want.

When you put particles in the cell then of course you get a lower radioactivity in the exit solution from the cell due to the fact that adsorption is occurring onto the particles. The lower curve in Fig. 8 shows the data that you get in that situation.

And at the end of the experiment, i. e. after 2hrs, you can rinse the system with a buffer and the radioactivity goes back down to zero again, showing that there is essentially no adsorption onto the tube passing through the detector. There is a technique that we have to use to prevent adsorption from occurring onto the tube walls.

Hoffman: It seems to me a lot of the interesting information is in the first few minutes when you're showing nothing coming out because the particles are sucking up the protein.

Brash: No that's not actually the case. The initial period of zero radioactivity is due to 'dead volume'; it takes the radioactivity a certain amount of time to reach the detector after you start injection since there is unmixed or 'dead volume' outside the cell which has to be traversed by the protein solution. But

182

after about 12 min, where the curves diverge, you do observe possibly complete depletion of the solution due to adsorption in the lower curve.

In that case of course you can't measure the rate of adsorption because the system is 'supply limited'. In other words everything going into the cell is being adsorbed.

Jennissen: Have you looked at the dependence of this curve on your stirring rate?

Brash: We have done some studies on that. It is somewhat dependent on stirring rate and one has to take that into account. If you don't want to be diffusion limited, as well as being supply limited, then you have to stir fast.

Jennissen: You have a grid on the outflow side. And actually this is very similar to our method (Ch. 8, Ref. 6) which dates back to 1976 where we also employ a grid for sampling. The problem is, of course a clogging of this grid or frit with particles or protein. Do you have problems like this in your system?

Brash: Well, as I said you're stirring vigorously in general, and you have to stir at least rapidly enough so that the particles are maintained in suspension and don't sediment back down onto the frit.

This stirring rate will depend on the particle concentration that you use. And clearly you have to have enough surface to deplete the solution significantly, just as you do in your measurement, Herbert.

The filter part of the cell is difficult, I wouldn't deny that; if there is an Achilles heel, that's what it is. And if you're working with particles of different sizes, you may have to develop a range of cells having a range of filter pore sizes to retain particles of different size distribution. But these are technical problems and I think they can be solved.

Wolf: :You keep the volume constant in your measuring cell, that means if you add from the one side radioactive solution on the other side the same volume is going out. Yes?

Brash: Yes.

Wolf: And that means practically you put the solution from the bottom and the overflow, is it such kind of overflow cell or how is it done?

Brash: It's literally as I described it in one of the first diagrams. It's a 'closed' cell if you want. It's just like a CSTR, the overflow is constant and the volume of fluid inside the cell is constant.

Wolf: And what about the relationship between the surface area in relation to the volume of your cell.

Brash: As I just mentioned you need to have enough surface.

Wolf: Yes, that is clear. But maybe you can tell us some values, maybe mls.

Brash: This is a cell of volume about 15 ml and the amount of surface in this experiment was 1000 cm². That's enough to give you a difference in radioactivity at the exit to the cell of the order of magnitude shown, which can be easily detected. You can do it with much less surface than that and still get data which are valid.

I just want to present a couple more figures on the adsorption data. You have to convert radioactivity in the exit stream into mass of protein adsorbed on the surface. When you convert the lower curve in Fig. 8 into mass of protein per unit area of surface, you get the kind of curve shown in Fig. 9. From zero to about 12 min is the dead volume time, as you pointed out Allan, and at this point you begin to measure adsorption.

The initial fibrinogen concentration that was injected was 95 μg/ml, and the volumetric flow rate 0.15 ml/min. The initial radioactivity was about 20,000 counts/min per ml.

Hoffman: John, you used the words 'dead volume time'. If you compare the blank without the particles with the one with the particles, you don't have that dead volume time at the one without particles.

Brash: You do.

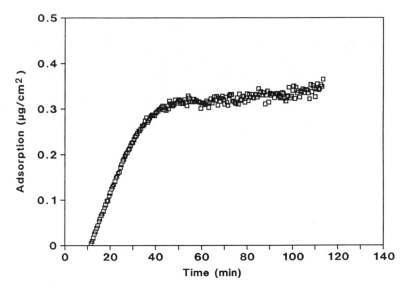

Fig. 9. Fibrinogen adsorption from isotonic Tris buffer (pH = 7.35), onto glass beads as a function of time. (C_{f0} = 95 μg/mL, v_0 = 0.15 mL/min, S = 1030 cm², A_0 = 11451 CP1/2M).

Hoffman: Yes, you're right. There is some. But it's different isn't it?

Brash: Yes, because at 12 min point adsorption can begin to be detected. It takes about 12 min from the time you begin to inject the solution until radioactivity reaches the detector, at the exit to the cell.

Hoffman: So, there are actually two dead volume times, in the sense, the first one...

Brash: You know that the radioactivity reaches the detector at 12 min, if there is no adsorption. Any difference between the two curves from this point on is due to adsorption.

I just want to point out that this is a technique which is reasonably good for measuring in situ adsorption with fast kinetics. There *is* adsorption in this case and you can take a time point as frequently as you want to, and you get values of adsorption which are in agreement with those in the literature.

So, Fig. 9 shows the kinetics. Adsorption increases rapidly onto a plateau for fibrinogen as a single protein.

Fig. 10 shows curves of adsorption versus time for diluted plasma injected into the cell. The plasma concentrations that we used were 0.1%, 0.5% and 1%. Actually there are two curves for 0.5%, which shows you the kind of reproducibility we can get with the technique.

These data show the Vroman effect when you go to high enough plasma concentration. Adsorption goes up and comes down again. At lower plasma concentrations there is loss of the Vroman effect because there are less of the high affinity displacing proteins.

Fig. 11 combines two independent experiments on the kinetics of adsorption of fibrinogen and IgG from plasma onto glass beads. Both curves show peaks, but the peak height is lower for IgG, and the peak occurs at an earlier time.

You asked me, Adam, two days ago I think, whether we had any data on other proteins besides fibrinogen. And that's the answer. And it fits well with the Vroman model from the point of view that the peak occurs for IgG at a shorter time than for fibrinogen. We hope to be able to do this with a number of other proteins, and see what the sequential order is going to be, in terms of time.

So, we're just beginning with this technique. It is not one that we invented, as I mentioned before, but we think is has a lot of potential. It has its limitations. For example, you can only use particles as the surface. You can't use flat films or tubes...

Baszkin: Can you use flat films? This is an important question.

Brash: If you can get enough surface into your cell, you can use anything, but from a practical standpoint it has to be particles to do that. Hollow fibers maybe.

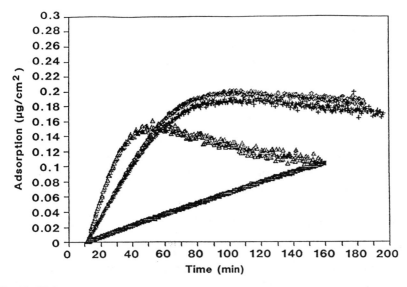

Fig. 10. Fibrinogen Adsorption from diluted plasma onto glass beads as a function of time. (v_0 = 0.15 mL/min, S = 685 cm²).
Injected plasma concentration:
□ = 0.1% plasma
+ = 0.5% plasma
   = 0.5% plasma
Δ = 1.0% plasma.

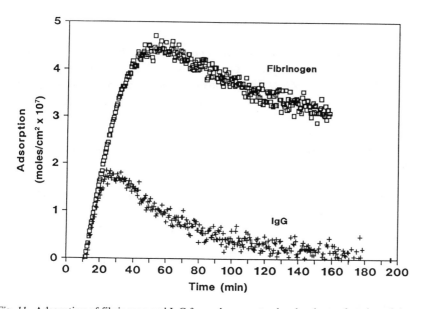

Fig. 11. Adsorption of fibrinogen and IgG from plasma onto glass beads as a function of time. Injected plasma concentration 1%.

Cazenave: How do you determine the surface of your particles? Is there a way to measure surface areas of particles of unknown surface?

Brash: You need to know the surface if you want to calculate the mass per unit area.

Cazenave: How do you determine that surface area?

Brash: Well, there are various ways to do it. You can do it by porosimetry, you can do it by nitrogen adsorption, and if you feel that's appropriate you can do it by adsorption of a protein of known properties, which many people do. You can do it by particle sizing, if the sizes of the particles...

Cazenave: What did you do in your case?

Brash: We worked with particles from a Company that told us what the size distribution was and that the particles were spherical. We looked at the particles in the microscope and verified this was the case. And we calculated the surface area from that.

Baszkin: This is a beautiful technique. However as all techniques based on depletion, protein adsorption can only be performed on developed surfaces. This means on beads, on powders, on particles, because you need a developed surface to have depletion.

Brash: By developed you mean large.

Baszkin: Yes, and my criticism concerns neither the technique nor the result. You cannot extrapolate protein adsorption data on lattices, powders, beads to the problems of biomaterials. The surfaces of beads and particles are not the same as those of finished surfaces produced out of these materials. So, you cannot screen by this method what you will obtain on finished surfaces.

Brash: What you're saying is somewhat speculative, and perhaps there is some evidence...,

Baszkin: This is what I'm repeating for many years and I would voluntarily like to put my name on this remark. You cannot extrapolate results obtained with powders, lattices, beads to the finished surfaces which are used later as biomaterials.
Brash: You may well be correct. I wouldn't dispute that point. But I'd like to see hard evidence on it. If you are correct, then we're left with being able to characterise adsorption onto materials which are in particle form, and will be used in particle form.

Hoffman: I think Adam has a good point. If you're talking about biomedical polymers that are commonly used, like poly(ethylene teraphthalate), or Dacron, that may be difficult.

However, when it comes down to some materials, silicons and Teflon for example, you can get surfaces that are quite characteristic of the finished materials in other forms.

John and I are going to be collaborating in our system with the plasma discharge deposited polymers, which don't care what the shape of the substrate is, it will come out the same composition.

So in some sense you can test it in a variety of forms with exactly the same chemistry. And particles will have it.

Jennissen: I think it's very interesting that you're working with particles now, as we have in these past years. I would therefore like to show some of the kinetic data we have obtained on our system (Ch. 8, Ref. 14).

Wolf: First question is: as far as I know adsorption kinetics it goes more fast than it's shown here in the results. Maybe also in your methods.

We have also used particles, glass beads and latices. We did find in terms of the change of electrical surface charge that we couldn't find, in case of adsorption for fibrinogen and HSA and other proteins a measurable effect after 2 min. That means after 2 min, which was the first possibility for us to measure we should say the surface was already covered by the protein.

And that seems to be fully different, so is there any explanation?

And the second question concerns also the method which I find very interesting and in case where you have a high stirred system of particles you produce a system as to make electrokinetic measurements. That means, in dependence on the speed of the particles in your suspension, you produce at the particle surface a different zeta-potential. That may be also influence the adsorption kinetics of such proteins.

And third question: if you make competitive measurements with your system here, then, of course, you have a system with a very complex hydrodynamics. That means you are in a situation if you are looking at one particle at the one side there is adsorption and at the other side, because of the movement of the particle in the fluid system, desorption. So it seems to be very complicated, or not?

Brash: With respect to the last question we are measuring the net adsorption, which consists of an adsorption term and a desorption term. And in the case of plasma you can see that at the early time phase of this interaction the adsorption in higher than the desorption; and in the later stages it's the reverse. The desorption is higher than the adsorption.

So the term for adsorption in that mass balance equation is a composite term for adsorption and desorption. You could write them separately if you want.

With respect to the fact that you feel the kinetics is too slow here in

188

comparison to what you have found, I think it would obviously depend on the concentrations of proteins that you're studying. The data in Fig. 11 for example are for an injected plasma concentration of 1%.

Now don't forget you have the volume of this cell which has no protein in it, and you inject 1% plasma into that. So at the beginning the concentration is perhaps 0.001% of what it would be in plasma. And it's increasing as a function of time. But it remains rather small.

The fact that the curve is bending over like that right from the beginning means that you are measuring a real adsorption rate. If the curve simply goes up linearly then you're probably supply limited. You're just measuring the rate of supply of protein to the cell, which then becomes a lower limit on the adsorption rate. The intrinsic adsorption rate can't be less than that but it could be more. Going back to the fibrinogen-plasma experiment, that was the case at 0.1% plasma (Fig. 10).

But I would say in comparison to your results, maybe it's just a question of concentration.

Zeta-potential: If we could subtract all the other effects that are going on in the system from the data that we get then we might be left with something, that we could interpret in terms of the charge on the surface which might be changing continuously as the experiment goes along. Is that correct?

Wolf: Yes, I think only if you change your stirring velocity, then of course you would produce a different surface charge at the same type of particles.

Brash: We should think about that. There may be a way in fact to extract information on zeta-potential and charge interactions.

Koutsoukos: I am concerned with types of methodology that involve separation of the particle suspensions from the fluid rate. And my concern is do you evaluate, or, can it be evaluated the adsorption on the separation level?

For example in your case you use a glass frit. Glass frit is expected to have a large surface area, maybe has different composition also from your glass beads. How much of the protein is adsorbed there? Or, do you assume or you measure that nothing is adsorbed, there?

Brash: We have a trick which prevents adsorption on to the glass frit. It's a technical detail and we verify that there is no adsorption onto the glass frit. You would either have to do that or else you would have to estimate the adsorption onto the frit. But it's an important point.

Cazenave: You have a high rate of stirring. Are you sure you don't get foam or bubbles into your system and what is the state of the protein in the effluent? Do you denature your fibrinogen, do you break it down, do you have dimmers or polymers, or do you loose functionality?

Brash: All we've done with fibrinogen so far is to take some protein at the exit from the cell and do SDS-polyacrylamide gel electrophoresis which looks normal. It looks the same as the gel before the protein went into the cell.

Cazenave: It clots?

Brash: That we haven't done. But we should. The radioactivity stays in the same position in the gel, the distribution of radioactivity doesn't change.

Baquey: I did not hear if the beads were somewhat porous. And in this case which were the permeation characteristic of the beads? I would like to know if you take small protein molecules. Did you have monotonous increase of the adsorption or are you able to see a double phase phenomenon, because of the access of the protein to the porous volume.

Brash: The beads that we used were non-porous.

Baquey: Not at all?

Brash: Not at all. You can look at them in microscope at high resolution. They are solid glass beads. At the level of a fibrinogen molecule, I wouldn't be quite as dogmatic as to say that, but they are non porous microscopically.

I think to use porous beads would be interesting and you could perhaps get some information regarding diffusivities and the like. But we haven't done that yet.

Missirlis: John, just a technical question. What is the final angular speed you have in your CSTR?

Brash: I'm only guessing. It might be 100 RPM or something like that.

Missirlis: So, it's not that fast.

Brash: No, it's not going to 'smash' things up too much. But it keeps the particles in suspension.

Jennissen: As I said, John, your method is very similar to ours except only that we take our sample through the grid on the syringe and we do not continuously add protein. It's a very nice method.

The following example shows you the power of our method in analyzing sorption kinetics (Ch. 8, Ref. 14). Here you see experiments analyzing the adsorption kinetics of phosphorylase b on butyl agarose. We measured the initial rates and it is possible with our method to obtain a resolution of ca. 10 seconds.

The experiments were performed at two temperatures 5°C and 30°C.

Plotting the initial rates versus the free protein concentration yielded hyperbolic curves which can be linearized in double reciprocal plots (Ch. 8, Ref. 14).

From this kinetic behaviour we have concluded that for phosphorylase there must be a rate-limiting step of adsorption similar to a rate limiting step of polymerization as has been found for the polymerization of tobacco mosaic virus.

So in comparison to John's method our method is based on discontinuous sampling to a grid. In our method protein is added only at the beginning with the sample volume only negligibly changing the constant volume. In contrast protein is continuously being added in John's method. Our time resolution is about 10 seconds.

Missirlis: So, coming back to the first question comparing static and dynamic data could one come back to say something on that?

Baquey: I think that it's quite difficult. You have to consider the same time scale. If you are sure that during the in situ experiments you are looking at the initial events, may be you may get some kind of comparison. But if you look at larger time scale you may get during ex vivo experiment different type of results.

Previously, we saw an accumulation of fibrinogen, or material generated from fibrinogen, related to variation in the flow condition, (Fig. 12), but I think that the reason for that is a different evolution of the thrombogenic phenomenon, instead of different initial events.

Jennissen: A phenomenon often encountered during the flow or stirring of fluids is unstirred layers (Ch. 8, Ref. 14). In kinetic experiments you either have to stir vigorously or need high flow rate in order to eliminate the unstirred layer. This can be easily checked by testing your system for a dependence of kinetic parameters on either flow or stirring rate.

Brash: Another way to look at the 'unstirred layer' question is simply that one must avoid diffusion limited situations in measuring interfacial events.

If, for example, you do an experiment which is too short, under the physical conditions that you're using, then you may only be measuring the diffusion rate of the protein to the surface, which would be the same for a whole range of surfaces.

By doing the experiment at high flow rates, or high shear rates in general, you tend to make the transport step faster and therefore reduce the risk of being diffusion or transport limited. So, I think it's a very good point.

I think one has to be aware of this problem and has to verify for a given experimental system, that one is measuring the adsorption event, that's to say the interaction between the protein and the surface, and not simply the transport rate.

Hoffman: The verification then would be: you're going to vary the shear rate in your system. In other words, your stirring rate. You'll come to some stirring rate where you don't see any further variation in the results.

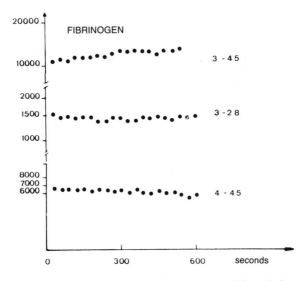

Fig. 12. Effect of the resistance brought by the extracorporeal loop during an ex-vivo experiment, on the behavior of fibrinogen. The extracorporeal loop is made of SCURASIL® tubing, the figures on the right referring to the diameter (3 mm) and to the length of the loop (45, 28 or 45 cm).

But eventually, you can go to a too high stirring rate where you begin (and Jean Pierre pointed out) to denature the protein, and then you should begin to see variations.

There should be some optimum stirring rate and that would be a deciding factor. Is that right?

Brash: You may not have to stir at all to not be diffusion limited depending on the conditions.

Hoffman: What about settling of your particles?

Brash: This would be a problem for cells but I don't think it would be for proteins.

Hoffman: The particle system is what we're talking about here.

Brash: Oh, yes, that has to be stirred.

Missirlis: Have you done, for example, your experiment at different RPMs and see how the data varies?

Brash: We haven't done a lot on that as yet. We are satisfied that the data we're showing are not transport limited in the case of this cell. But I think Allan is right that a possible way to avoid this problem is to do some preliminary experiments.

It's a little bit like using labelled proteins, verifying that the label doesn't influence the adsorption of the protein.

At the beginning, before you go too far with your experimental system you should make some kind of test that you're in fact measuring adsorption events and not transport events. And one way to do it is to vary the flow conditions, and show that your measurement is independent of that.

Wolf: What is the main difference between your method or this method and 'column bead method'? That means that you fill particles like Prof. Jennissen in a column. What is the real difference in your method?

Jennissen: We're not using the column, sorry.

Brash: It's not a column. It's a well stirred column if you want.

Wolf: Yes, but if you take a column filled with particles then you have not this equilibration, what you mean, by stirring. That's the advantage.

Hoffman: But, basically, this is a system that eliminates the problems of chromatographic separation. You're doing adsorption chromatography in a suspension rather than in a column where you avoid the channelling, the radial diffusion effects, and so on, and here you have a very good mixing. So, you avoid the problem of diffusion and you're looking at the adsorption event.

Koutsoukos: One point also I would like to make the experimental approach is that is not only flow conditions that one should pay attention especially if one works with particles. But also is the size of the particles. Because these two combined both, flow conditions and particle size determine the thickness of the diffused layer around each particle.

So, if say there are particles that are below 0.5 μm or about that size and the flow rate is in an intermediate range then one may end up with increased concentration, higher concentration around the particle than in the bulk solution. And also not only the concentration of the protein molecules maybe higher, but also the pH may be quite different from what is in the bulk. A factor which would be very important in conformational changes of the protein molecules.

Jennissen: In our system sorption kinetics (Ch. 8, Ref. 14) were independent of the stirring rate (2 cm stirring bar) between 200 and 1200 RPM. Below 200 RPM we would get lower rates. Our data did not go through a maximum but reached a plateau above 200 RPM.

Wolf: There is no foam formation then?

Jennissen: If you go higher that 1200 RPM you get foam formation.

Brash: There is another way to check on diffusion limitations which is to do a calculation of the concentration profile by solving the diffusion equation for the geometry that you have.

Now, this is not always possible if the geometry is too complicated but it can be done for simple geometry like, for example, flow through a tube.

Then if you find that the concentration is 0 at the surface then you're still diffusion limited at that point.

If you calculate that your wall concentration is above 0, then you're probably not diffusion limited. But I would caution it's not always that easy to solve the diffusion equation for a given geometry.

Missirlis: Can I say something else here? How can one relate data of adsorption of a single protein on beads, on flat surfaces of different flow rates and on static conditions? Can you relate all this?

Baszkin: To my personal opinion and according to the results obtained in my laboratory, this problem does not concern only hydrophobic polymers like silicons, as you have mentioned Alan. It concerns also polymers having the surface energies higher than 40 or 45 mN/m. It is very dangerous to extrapolate every possible results obtained on powders to biomaterials which are not powders.

Depletion methods are very sensitive, but please don't show them to people who produce polymers for biomedical uses because they will think that they have a useful method for screening their polymers.

Hoffman: Well, it's not my method, so I'm not going to take issue with Adam.

I still claim that you may be able to study the protein adsorption kinetics in a dynamic sense in these systems. Surfaces that are well defined and characterized by ESCA are therefore extrapolatable. You can also characterize surfaces, whether they're very polar and hydrophilic or non-polar, using ESCA. I think this can be used for those comparisons. I don't see any problems with this.

This question of static versus dynamic is a whole another issue. This does not lend itself to static studies because the particles settle and you might as well use a flat surface.

I had one other question. Is it worthwhile to avoid the foaming to degas your buffers before you use this system?

Jennissen: Under our conditions foaming was only observed at very high stirring rates. We did no t have to degas our solutions. When working at higher temperatures all solutions and the adsorbent (gel) were prequilibrated to the new temperature before mixing and spontaneously degassed so we never had any foaming.

We've only talked about methodology up to now. We haven't talked about the rates and mechanisms; will there be time for that?

Missirlis: Yes, we could continue on after the break.

Missirlis: There are some questions how relevant and valid are experiments on surfaces that are not used in this geometry, say, in the actual application and how one can extrapolate or relate data on two different methods and, of course, the dynamic versus static experiments and the actual flow conditions that would be found in the final application. So, if you please comment on this.

Lemm: Or, more precisely, how do you transfer you RPM in your beaker to the flow for example in the venous system.

Cazenave: Well, I think I'll try to give a general comment and relate that to what we are faced with in the development of drugs for human use. You need all the steps, you need to do experimental animals, you need different shapes of polymers, tubing and beads and then the final test if it's a biomaterial for human application is man.

So, I don't think that you should say this type of test on beads on powder is useless. I think it has its own limitations but it's an easy way to go, because we cannot go directly to the final product in the final shape, in the final form to be implanted in man or in animals. So, I think there is no irrelevance.

Of course some screening tests might be fast to do, easier to do, cheaper and closer to the human use, but sometimes you have to do simple things if you want to screen large quantities of materials. This has been done with drugs. And still some drugs go to development in clinical phase one or two, and have to be withdrawn because there is a side effect which could not be predicted at all if not used in man.

Lemm: Do we all agree that the results under static conditions differ from those under dynamic conditions, and the protein composition on a surface?

Brash: I think in general one has to check. It's not axiomatic that the adsorption is going to be different under flow versus static conditions. That may well be but it doesn't have to be.

You can imagine certain conditions in a static experiment where you're not diffusion limited. In that case flow may not really affect things very much.

I would make one point in this regard, namely that as we go closer to the real situation, what we have is blood flowing past the surface, containing cells and particularly red cells. These are known to really stir things up at the local level and by convective effects to 'throw' packets of fluid, at the surface.

This is an old and well studied effect. Jean-Pierre Cazenave has studied it a lot, and we have too. I'm not so sure if that has been well established in the case of proteins. But in the case of smaller cells like platelets, it's absolutely crucial that you have red cells present in your system.

May be somebody should investigate it a little bit more with respect to proteins. I think that may well be important in the final analysis. If you really want to know protein adsorption kinetics in blood, at least have red cells, or some particle which is similar in shape and size to red cells, in your system.

Hoffman: I recall John Andrade made a list of what he called the key 12 proteins in plasma, and he characterised their effective diffusion constants as a function of their size and so on. And the rate at which they would approach the surface in static diffusion.

So, I think that there will be always a difference between the static and dynamic case. In that sense John I would disagree that I think there are maybe key proteins that are large that do take time to get to the surface. That would potentially change the kinetics, the dynamics of the result in cellular interactions, if it was done in static versus flow system.

But you brought up a good point about the red cells. They are well known to cause disagreggation of platelets near the surface and so in the absence of both red cells and platelets you can fool yourself about the interaction of the protein coated surface with whole blood. So that's very important, you're right.

Jennissen: I think the basic mechanisms of adsorption will not be so much different in a static and/or dynamic system. The kinetics will be very much different (e. g. unstirred layers). I think that's the point.

Brash: I believe that kinetics at very short times is probably always going to be influenced by flow, in the sense that the protein has to get to the surface somehow and it will get there much faster under any conditions, than under static.

However many people are interested in measuring isotherms to get thermodynamic data, particularly affinities, in which case you may well not have to worry that much about the flow conditions since you have to make such measurements at long times. But we are talking about kinetics here, so, let's be clear on that.

Beugeling: I fully agree with what has been said about the red cells. As you know, in the Cazenave's system, Cazenave as well as we have a suspension of platelets and red cells. Just for the same reason you mentioned. The red cells are necessary to push the platelets to the wall.

Magnani: I think also that FT-IR spectroscopy should be considered for adsorption and desorption studies in static and dynamic conditions. What is the primary advantage in the use of this technique? Of course the response time. The time necessary to collect data is very short: about 1 sec. That allows us to obtain information about the real initial events that occur when blood comes in contact with a given surface. But can I show how the data are collected?

Cazenave: To come back to what Tom said, we have done some experiments not with this capillary system but with the previous system we had, with another cell and we showed an influence of shear rate on fibrinogen adsorption to a glass surface, in the presence of red cells, but not on albumin adsorption (Ch. 6, Ref. 15).

By increasing shear rate in the presence of red cells we increased adsorption of fibrinogen on the surface and we had the idea that it might be due to the size and the shape of the molecule.

Brash: Can I just clarify? Was this adsorption *rates* that you were studying? Was there a difference in adsorption *rates*?

Cazenave: Yes.

Magnani: This is the ATR flow cell for adsorption/desorption measurements (see Ch. 10, Fig. 3).

The ATR crystal (generally Ge) can be coated with the tested polymer and every second, during the single or double protein solutions or plasma or whole blood flowing through the cell, data can be collected. By plotting then, the absorbance of the normalized Amide II-band, which is sensitive to the amount of the adsorbed protein, as a function of time, information about the adsorption isotherm can be obtained.

The same item can be performed in the desorption conditions obtaining, at the end, information on the amount of the protein irreversibly bound to the surface of the tested polymer.

I haven't any experimental data to show you, because up to now I have just performed static measurements for single protein solution.

Hoffman: How thin can you deposit your polymer coating? It must be very thin.

Magnani: The thickness of the coating polymer has to be less than the penetration depth of the IR beam, since the IR beam has to go through the full thickness of the coating polymer before to see the protein which is adsorbed onto its surface. The penetration depth in the case of Germanium is about 0.2-0.3 μm, so that the coating polymer thickness has to be about 0.1-0.2 μm or less.

Brash: I think we may have touched on this before and you may have mentioned it in your presentation. I refer to the problem of subtracting the bulk IR signal. This is a potential problem with that kind of total internal reflection method and everybody who uses it has to face it. And I'm not convinced that people have entirely solved it, but maybe you can tell us.

Magnani: Yes, in ATR experiments performed with physiological concentrations of proteins, the IR signal is composed of contribution from the adsorbed proteins and from soluble proteins in the 'bulk' solution. The procedure for estimating the magnitude of the bulk contribution and correcting for this effect is not so easy to perform.

Anyway a 'two-cycle approach for separating the adsorbing protein features from those of the protein still in the liquid layer near the surface was investigated by K.K. Chittur et al. (Ch. 10, Ref. 4). In this approach, a clean surface is exposed to a flowing protein solution for a period of 1-2 hours, to produce a stable protein film on the surface; the cell is then washed with buffer to clean the protein in the supernatant and to yield an IR spectrum of the adsorbed film.

In the second cycle more protein is perfused through the same cell under the same conditions of the cycle 1. Subtraction of the spectrum of the adsorbed protein film from each spectrum of cycle 2 yields a series of spectra that describes the filling of the flow cell in the absence of protein adsorption to the surface and that can be used to approximate the development of the bulk even when adsorption does occur.

Subtraction of this series of 'bulk' spectra from those in cycle 1, at the corresponding times after introduction of protein to the flow cell, results in a series of spectra that approximate the build-up of the adsorbed protein film.

Missirlis: John could you answer the question of how you transfer, say, the information from RPM to flow conditions in the venous system?

Lemm: For example. Or in an artificial lung or kidney?

Brash: I'm not an expert on fluid mechanics, and I think somebody who is ought to answer that question. If there is one here may be we should hear from him or her, but I'm pretty sure that one can do calculations and fluid mechanical analysis using fairly standard, well known procedures that would enable you to calculate the average shear rate at the surface of the particle.

Obviously you would have to use an average when you have a large number of particles with a distribution of velocities, but I think calculations of that type can be done and then compared to shear rates in typical venous flow or arterial flow situations.

Lemm: About ten years ago, I made an experiment : I simply circulated a protein solution containing albumin and fibrinogen in competition through a tube with an inner diameter of 4mm. With increasing flow I found that the preference for fibrinogen increased too. How can this be explained?

Brash: It sounds to me as if you may have been diffusion limited or at least transport limited from the fact that your preferentiality for fibrinogen increased with flow. As you increased the flow you increased the transport rate.

The effect of flow on the fibrinogen was greater than it was on the albumin. It's a possible explanation.

Missirlis: Because they have different sizes.

Brash: Very different sizes.

Jennissen: Yes, it may be a problem of shape. Dr Lemm's experiment reminds me of a flow phenomenon observed with actomyosin molecules. If you have actomyosin molecules which have a very large length to width ratio (i. e. it's a very long molecule just like fibrinogen) flowing e. g. through a tube they will align according to the flow and you can get a so-called flow birefrigence because the molecules become ordered.

Maybe in your system you're aligning the fibrinogen molecules in a flow dependant manner so that it will preferentially bind to your tube. In the alligned form it may be capable of making more contacts with the surface than when it is tumbling at very low flow rates. It's just a speculation.

Lemm: So what can we learn from this? We should have a lower flow for example in...

Jennissen: To eliminate unstirred layers you usually need turbulent flow. This would however also eliminate flow alignment. If you're having a laminar flow...

Lemm: It was turbulent.

Jennissen: Well maybe not on the edge or close to the surface.

Hoffman: I think in the end you have to use shear rates that are relevant to blood flow through the body and through devices. If you're going to turbulent flow, this is generally not encountered in the fluid mechanics of blood, at least whole blood, in devices or in the body. So, I think that's something to avoid.

Other things that can happen is, if you increase the shear rate too much you can begin to denature proteins and you can get that preferential adsorption due to that. Also potentially you can shear off adsorbed proteins which adhere very strongly such as albumin. All of it is very complicated. So it's an important variable.

Cazenave: Do you think that there is not turbulent flow in the body? I think there is a lot of turbulent flow in the body but engineers like to have flow with equations. Laminar flow, it's easy to understand.

Hoffman: Yes, no doubt there are some very irregular flow patterns coming out of the aortic valve near the heart. In the bifurcations in the arterial branches where the high pressure is, you get flow separations due to deposits of fatty plaque and so on.

These kinds of flow separations can still be and probably are stream line or non-turbulent flow. It's a flow separation, but it's not turbulence in the sense of the Re number, that type of turbulence.

Beugeling: I'm also not a medical doctor, but that what you mentioned, I think is not true, because these are just the places where thrombosis begins.

Missirlis: But, let's not swim in very deep waters outside our system here.

Hoffman: Let me just caution you I don't want to go on record now. I will correct the record when I have a chance. I'm saying that there is no turbulence in the body because the word 'turbulence' is so ambiguous. I'm not going to argue about that.

Missirlis: Of course, flows in the bodies are neither extremely laminar nor extremely turbulent. We have flow separations. it's a whole mosaic of different types of flows. And it's a little bit outside, I think for our theme.

18. The Vroman effect

Missirlis: But could we also just move a little bit: somebody should define the Vroman effect and define the ultrashort time tests. What do we mean by ultra-short time and obviously, in my mind, it seems it's the most important thing, just the initial contact and are there tests, and what we do about it and, how then we correlate these tests to the more conventional ones?

Beugeling: I want to make a comment on the Vroman effect and I give a definition which has been given by André Poot in his thesis (Ch. 4, Ref. 10).

'The Vroman effect is characterised by a decrease of the amount of the initially adsorbed fibrinogen from plasma onto foreign surfaces with increasing contact time, as well as by a maximum in the adsorption of fibrinogen as a function of the plasma dilution.'

I think this is in general what happens. I want to add that you cannot exchange time for plasma dilution, because in the case of diluted plasma the surface is not fully occupied with adsorbed proteins in the beginning, so proteins which adsorb have the possibility to spread out, and that is less when you have concentrated plasma. So these two Vroman effects are not the same.

What has been said for fibrinogen is also true for many other proteins. In the case of fibrinogen it's found for glass and glass-like surfaces. But in the case of polymers it's totally different. I already mentioned that, in our opinion, HDL is involved in the Vroman effect.

Brash: I think that definition of the Vroman effect is somewhat limited. You defined it in terms of fibrinogen which is a natural thing to do because that was how the Vroman effect was first brought to light. But people are thinking of the Vroman effect at the present moment in a more general sense. The key word here is sequence: the proteins appear on the surface in a sequence which is as yet not well defined, but there is a sequence and it's normal that there should be. Because of 'concentration times the square root of diffusion coefficient' effects.

Alan referred to this with respect to Joe Andrade's analysis. The more abundant proteins with higher diffusion coefficents, i. e. the smaller ones, will

Y.F. Missirlis and W. Lemm (eds), Modern Aspects of Protein Adsorption on Biomaterials, 201-217.
© 1991 *Kluwer Academic Publishers. Printed in the Netherlands.*

move to the surface more rapidly than those with smaller diffusivities and lower concentrations. So they'll get there first and they'll adsorb.

Later on as the slower proteins move to the surface they'll be able to exchange with the initial proteins, if they a have higher binding affinity for the surface.

That is, in my mind at least, the essence of the Vroman effect. It just happens that it was first noticed for fibrinogen. But I think there is a fair bit of evidence now that it happens with many other proteins, possibly all proteins.

It's not a question, however, of 'black and white' where at some point in time the surface is covered with albumin and then with IgG and so on down the line. I don't believe it's as clear as that. But there is an element of sequentiality in the time evolution of proteins on the surface.

I think we ought to discuss this apparent discrepancy that you seem to have found regarding the fact you don't see the Vroman effect as a function of time. Because we always do.

Beugeling: I didn't say that. It is a function of time, of course. If you look at concentrated plasma, then you see a decrease of the amount of initially adsorbed fibrinogen and of other proteins too.

The second Vroman effect is the dilution effect in which there is maximum of the amount of adsorbed protein. So there are two things. But these two things are different.

Jennissen: In the Vroman effect you probably have quite a number of effects going on at the same time. I would like to present some figures (Ch. 8, Ref. 14) on the cooperative displacement of protein from a surface because they may be the mechanistic basis of the so-called Vroman effect.

I'll describe the system. What we do is we isolate the gel with the adsorbed radiolabelled protein and then redilute it. We dilute it once in buffer containing no other protein and then we dilute it in buffer containing cold labelled protein.

And what we measure are the initial desorption rates. The initial desorption rates are linear in a semi-logarithmic plot indicating that we are working under first order conditions or pseudo first order conditions. Therefore we can derive an off-rate constant from these data.

Here you see that we obtain a four-fold lower off-rate constant in the absence of unlabelled bulk protein as compared to the off-rate constant in the presence of bulf protein. This effect is independent of dilution.

If you were having readsorption of labelled molecules, then you should see a dependence on dilution. This shows you the independence of this effect on the stirring velocity. Ch. 8, Ref. 14, Fig. 6A shows you an increase in the off-rate constant as a function of the initial bulk enzyme concentration.

We have here the control in the absence of bulk enzyme and here we add 1 μM concentrations of bulk enzyme, 10 μM and 100 μM and you see that the factor difference increases from 1, 3 fold to about 5 fold, which indicates that the concentration of the bulk protein which is displacing the labelled protein is decisive in the effect you obtain.

And the bottom (Ch. 8, Ref. 14, Fig. 6B) shows you the dependance of this effect on the surface concentration of butyl residues. And we have very interesting finding here: at a very low concentration of butyl residues, that is when you have a low affinity of binding, there is practically no effect between the desorption in the buffer and in the presence of enzyme. As you increase the surface concentration this effect appears. You see here we have about a 4 fold effect here at 0.45 μmoles/m^2 of immobilized butyl residues. So, this effect depends strongly also on the affinity between the protein and the surface.

And finally we looked at the dependence of this effect on the fractional saturation of the surface, and that is shown here (Ch. 8, Ref. 14, Fig. 7). And you see that as we increase the fractional saturation from 0-0.75 the difference between the dilution in the absence and presence disappears, which means that only at fractional surface concentrations in a quite low range you can see this effect. We interpret this to mean that there has to be a space between the adsorbed molecules for the unlabelled molecules to be adsorbed in, in order to be able to displace these molecules from the surface.

Van Damme: I would like to comment on the comparison between the concentration domain and the time domain.

In the experiments we did at polymer surfaces, we always showed the Vroman like curves in the concentration domain so we always observed a maximum.

But when we looked at the time domain we never observed a decrease of any protein. For some proteins we saw an increase like HDLP but we never a decreasing amount of protein on the surfaces was observed.

So I wonder if the comparison between the concentration domain and the time domain is so logical. Of course it's a different. You're looking at two different things.

Also, I think in literature there are some reports when you look at the time domain of a single protein solution that you see a decrease in the amount of proteins adsorbed. There is at least one other process of desorption of proteins. They can't be exchanged in single protein solution.

Hoffman: First, I had a quick question, and then I wanted to show two slides. What was your time domain? Maybe it was too long.

Van Damme: The time domain we worked was started at 15 sec up to 20 hrs and the concentration domain we started at 100% plasma up to 10^5 dilution.

Brash: Along the same lines what time did you use when you did the concentration domain curves?

Van Damme: 1 hr.

Brash: Did you use different times to do concentration domain curves?

Van Damme: No, only 1 hr.

Brash: You may miss some effects by working only at one time, you may have to vary the time.

Van Damme: But in the concentration domain we saw the effect very clear. And that's a bit surprising because if you say it's something from the methods then you would expect that you don't see it in concentration domain and not in the time domain.

Brash: Well, if it is the Vroman effect that you're looking at, then logically, and if the explanation that some of us have for this effect is correct, the time domain and the concentration domain should be superimposable not only observed but related to each other.

My definition of the Vroman effect is that proteins follow each other in a sequence to the surface and they replace each other as far as they can, in terms of their surface affinity. And this is demonstrated by a peak in either the curve with respect to time for a fixed concentration or the curve with respect to concentration for a fixed time.

Jennissen: In the time in between can I ask a John a question on the mechanism of the Vroman effect? You showed in one of your papers that proteolysis was going on in the same time that the Vroman effect is being measured. Could it be that the Vroman effect is not just a kinetic adsorption-desorption effect but is an expression of the kinetics of a limited proteolysis reaction on the surface?

Brash: I think one has to worry somewhat about proteolysis particularly in the case of fibrinogen where there is the possibility of plasmin generation in the system. Some of the experiments that we did had this in mind.

The time frames of the experiments were such that the possible proteolysis that could have taken place in that time would be rather small compared to the Vroman effect that we were able to measure. Another way to proceed is to put inhibitors of proteolysis into your plasma.

Cazenave: Can you show it with a mixture of proteins which have no enzymatic activity? If you take purified proteins, let's say 5 or 6, and you mix them, none of them being a substrate for the enzymatic activity of another, can you see the Vroman effect?

Hoffman: I'll just take a couple of minutes. First, I just want to point out one definition or one way to look at the Vroman effect as far as we are considering it. John Brash just described it and I think this is essentially what it's saying. You have two parameters of time and concentration. If you look at the adsorbed protein per unit area as a function of log of the% plasma concentration for adsorption of any one protein, then you see there is a maximum at a particular plasma concentration or dilution of the adsorption, and that maximum tends to shift to the left with time.

This is one way to look at the Vroman effect. I'll leave it at this point. Briefly, I wanted to show the Vroman effect can vary with the surface composition to a point where you may not even see it.

These are some data that we took on poly(ethylene teraphthalate) films using baboon plasma. I think it was 80% plasma. Tom Horbett and I worked together on this with his technician. This is the amount adsorbed as a function of time of the fibrinogen. You see the typical data that you get: 15 sec was the first point. You can't even begin to see the buildup to the maximum, but it's there.

Here, you see the 0 point on the axis. You can see we come up to a fairly high level and then it decreases as it's displaced by other proteins in the plasma and or some other effects of that surface.

On the other hand, when you treat the surface in the plasma gas discharge with a fluorocarbon coating at the same scales you don't see the initial peak buildup. If it's happening, it's happening in the first 15 sec. Or as Henk said, maybe you don't see it at all, for reasons that we don't understand. But the important point is this goes along with the second topic on your discussion of the very initial events.

If they're less than 15 sec, then they are different, depending on surface composition. We see many biologic consequences in terms of thrombogenicity which fits in with tonight's discussion. So, that's the point.

Cazenave: How do you measure fibrinogen?

Hoffman: ^{125}I.

Jennissen: Would you say that what's happening in the Vroman effect is an increase in the off-rate for fibrinogen?

Hoffman: Increase in the off-rate for fibrinogen? Do you mean the Vroman effect definition or with respect to this particular compositional effect?

Jennissen: Well you show that the fibrinogen came off the surface the adsorption rate constant should actually remain the same. So it must be an increase in the off-rate.

Hoffman: Maybe the amount that's adsorbing is different and it's not coming off the surface at all. Maybe what goes down here sticks and what we saw coming off earlier was not so strongly held. But that was on a different surface. So, maybe the Vroman effect doesn't exist on this surface. That's the question this raises.

Brash: I think it's more than just a question of off-rate. You can consider the off-rate to be some kind of natural desorption rate or in addition that it's an effect where there is displacement of one protein by another.

I believe that's what's happening in the Vroman effect in plasma. But I think your mechanism for the effect to happen at all is undoubtedly correct.

Hoffman: You agree that it might not be happening in the second set of data?

Brash: It looks to me as if there is rather a lot of scatter in that data; I don't want to be unkind but I'm not really sure. I would draw a horizontal line around 0.07 or something like that.

Hoffman: Let's see again the previous picture.

Brash: I would say that there is no peak in the data. There could be a peak between 0 and your first measurement, whatever time that is.

Hoffman: 15 sec.

Brash: O.K. there is a peak, and you don't know how high it is because you can't see it. It may well be ten times as high as your first point.

Hoffman: Yes, you're right

Brash: And there is undoubtedly a peak in the adsorption as a function of time on that surface.

Hoffman: Is it possible the Vroman effect doesn't exist for some surfaces?

Brash: Yes, I believe that's the case. I presented a diagram previously (Ch. 5, Fig. 6) of the sulphonated PU surfaces but the adsorption is high and the concentration domain Vroman effect looks like an isotherm. It looks like a Langmuir isotherm and levels off in a plateau.
   So, yes, there are such surfaces. At the beginning of this work I used to think the Vroman effect was something that happens on every surface, but with different parameters: different rates of displacement etc. But there is no question that there are surfaces on which it doesn't happen.

Hoffman: I think that's an important point because the impression has been in a lot of minds that the Vroman effect is an absolute universal parameter.

Missirlis: What sort of surfaces don't show the Vroman effect?

Cazenave: The Vroman effect may not appear for your second surface as far as fibrinogen is concerned. If I understand, well, it depends on what kind of molecular labelling or antibodies you have to look for the 250 proteins which are in plasma.

Brash: That's exactly right. That goes back to the definition of the Vroman effect. I think we all tend to entertain a too narrow definition of it. We should say the 'fibrinogen' Vroman effect. Lack of a fibrinogen Vroman effect may be explained in the sense that maybe the sequence stops at IgG and the fibrinogen cannot displace it.

Jennissen: I think you have to go down to the basic mechanisms and define the Vroman effect in terms of on- and off-rates. Only then you can make comparisons. You need to measure the off-rates of normally adsorbed fibrinogen and compare them to the off-rates encountered under Vroman effect conditions.

Beugeling: But that depends on protein concentration.

Jennissen: That depends on many factors.

Beugeling: I fully agree with you. But then you have to measure it under circumstances in which you actually see the Vroman effect (plasma). If you are using protein solutions, you have a totally different situation. If the solution is diluted the adsorbed protein has the opportunity to spread out on the surface, and this is less if the protein has been adsorbed from plasma.

Baszkin: I have demonstrated that for many types of surfaces there is an important fraction of what we call loosely bound protein.
I would like to draw your attention to the fact that Vroman effect may be different on the surfaces before the rinsing and the surfaces after rinsing.

Van Damme: I've a short question to Profs. Hoffman and Brash. Did you ever measured Vroman effect without using ^{125}I and did you check if what is coming off the surface is really protein bound iodine?

Hoffman: I think we've always used ^{125}I label, although recently we were looking with ELISA at the protein on the surface, but that has to do with its conformational state. No, I think we were always using [125]I.

Brash: We've done Vroman effect studies at long times in the concentration domain using just conventional depletion techniques such as Dr. Jennissen uses and we've been able to show that it happens. The effect can be demonstrated with that kind of data as well.
With respect to the retention of the label by the protein, which I think was your second point, our typical experiment is for 5 min. And in that time certainly there is no significant loss of label by the protein.
We haven't checked at longer times although we have done that kind of check in other experiments and have been able to show that most of the label is retained by the protein.

208

Van Damme: But that's only in solution then. In the Vroman effect you first adsorb the labelled protein then get it off again.

Brash: May be I'm not understanding your question. Can you just repeat?

Van Damme: Did you check if the desorbed fibrinogen, which you measured was still iodine coupled to a fibrinogen molecule. If, you look at the Vroman effect, you see a decrease in protein concentration on the surface which means it desorbs into the solution.

Brash: Well, but that would be difficult to do, I think. Because you have a small amount of labelled protein which has been on the surface and come off and most of the protein that has never seen the surface and has not done anything so I'm not really sure how that experiment can be done.

Lemm: Is it of interest to have precise information within an ultra short time period (between 5 and 10 seconds) about the real initial event?

Hoffman: As I pointed out with our data, not very many people have correlated longer term effects with the initial first few seconds events with protein adsorption.

This is a very good question, because, it's more of a question that anybody can answer. It has been a premise, in this business that protein adsorption is an initial event which leads to cell interaction, such as thrombosis, complement activation, neutropenia and all the other events that we see with the contact of blood with devices and implants. The correlations are almost non-existing. At least we have seen, with that data I showed you, significant differences in those two surfaces in contact with blood in an ex vivo shunt.

Whether it's related to those differences or not it remains the question.

Jennissen: You could answer this question quite easily John. Just perform a trichloroacetic acid (TCA) precipitation of your fibrinogen. What remains in the supernatant are non-protein iodine or non-precipitatable (proteolytically nicked) peptides. What is precipicated is fibrinogen.

You could analyze the protein precipicated by electrophoresis in the presence of SDS and make an autoradiogram to see if the radioactivity is in the subunits of fibrinogen.

So I think the open question on what you are actually desorbing could be answered quite easily.

Turning to another question. Have you ever tried to simulate the Vroman effect by a sequential method, first adsorbing pure fibrinogen and then adding serum proteins. Does the fibrinogen come off?

Brash: That was one of the first things that we did. In a paper that we published before the Vroman effect, before we even knew about it, we put fibrinogen on the surface and then exposed it to plasma (S. Uniyal and J.L. Brash, *Thromb. Haemostas.* 47, 285, 1982). The fibrinogen came off rapidly.

What stimulated us to find the Vroman effect was that, when we did experiments with plasma, we didn't find any fibrinogen on the surface. If you did it with 100% (i. e. undiluted) plasma you couldn't find any fibrinogen on the surface.

Yet, if you did an analogous experiment with mixtures of fibrinogen and albumin, with the albumin in great excess, you saw mostly fibrinogen on the surface.

So what was happening to the fibrinogen? That was the question. I was stimulated by that plasma experiment, which showed no fibrinogen on the surface.

So what we did after that was to precoat fibrinogen onto various surfaces and then expose them to plasma. And the fibrinogen was gone in 5 min or less. With initially a close packed monolayer of fibrinogen on glass or even PE, the fibrinogen is gone in a few minutes after plasma contact.

Missirlis: Let's see some of the slides of Prof. Beugeling and then answer the question what happens in the first few seconds and what is the time limited thing, you know: how fast we can measure?

Lemm: Or more precisely, did really someone make experiments within the first 5 seconds?

Beugeling: No, we have measured from 15 seconds on.

Let me first say that all these measurements are made by the two-step EIA. The surfaces are rinsed with buffer after plasma or plasma solution has been in contact with them. Therefore we are not talking about loosely bound protein in terms of Prof. Baszkin.

These curves give the adsorbed amounts of fibrinogen, HDL and HMWK from 1:1 diluted plasma to glass as a function of time (Ch. 4, Fig. 5).

The reason that we take 1:1 diluted plasma is the following: we first add 200 μl of buffer into the wells of the test cell and then 200 μl of plasma with the pipette tip under the surface of the liquid. Then we gently stir it in order to get our solution. So we have no air-solid-plasma interface.

You can clearly see the Vroman effect in the case of fibrinogen which is not surprising. In the beginning the adsorption must be higher. We also see the Vroman effect in the case of HDL. You don't see it for HMWK.

In this connection I also want to mention that we don't see an increase of the amount of adsorbed HMWK in the first few minutes and that means that HMWK is not the only compound which is involved in he displacement of fibrinogen from the surface.

In Ch. 4, Fig. 6 you see the adsorption of these proteins and other ones to PE as a function of plasma dilution. The exposure time was 1 hr and then you see the second Vroman effect, the maximum in the amount of adsorbed protein as a function of plasma dilution. But I want to say again, this one is different compared to the one in concentrated plasma solutions or plasma itself, because in this case the solution is so much diluted ($\pm 10^3$) that the surface is not fully occupied by proteins in the beginning. Proteins which adsorb have the opportunity to spread out and to change into a state in which it is more difficult to remove them from the surface by other proteins. And in the case of plasma there is a large competition between the various proteins. The opportunity to spread out is diminished.

I already mentioned the adsorption plateau for HDL. HDL is preferentially adsorbed to such a hydrophobic surface as PE. The relatively low adsorption value of HDL is due to the fact that HDL is big molecule and only parts of it are apoprotein-A1 to which the antibodies are directed. This results in a relatively small surface concentration of the antibodies.

HMWK itself also shows the Vroman effect if it is adsorbed to a hydrophobic surface (Ch. 4, Fig. 6). You see HMWK is also decreasing at high plasma concentration, which means that it is also displaced. The fact that there is a rather high value at a plasma dilution of $1:10^3$ is only due to the activity of the enzyme-labelled second antibody, which was very high. I've already said before: you can not compare 'measured amounts' of different proteins which are adsorbed to the surface.

Missirlis: Tom, that's very clear. You said your first measurement was done at 15 sec.

Beugeling: Yes, after 15 seconds you see a typical Vroman effect, also for HMWK. Which means that it's displaced from the surface, if it is a hydrophobic surface.

Missirlis: Could you make measurements below that?

Beugeling: No, that is not possible with our method.

Missirlis: Could you think of a method to make the measurements say at 5 sec?

Beugeling: Not, not with the EIA. It's perhaps possible with the reflectometer if you use a flow cell. In principle it's also possible with the EIA but that's rather complicated and it takes a lot of time because you have to do so many handlings with the EIA.

I want to say something else in relation to the Vroman effect. When we observed a decrease of the amound of the adsorbed fibrinogen in the case of concentrated plasma solutions, we wanted to know which component was responsible for it. And we did competition experiments in which we measured the adsorption of a protein from protein mixtures.

We measured adsorption from mixtures of fibrinogen and very low density lipoprotein (VLDL), fibrinogen-LDL, and fibrinogen-HDL. The amount of

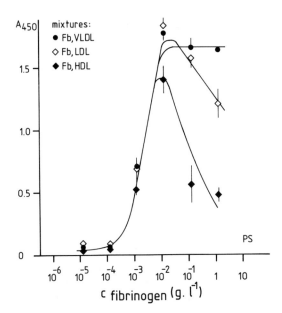

Fig. 1. Influence of lipoproteins on the adsorption of fibrinogen (Fb) to polystyrene (PS). (Fb : LDL = 4,2:1 w/w and Fb : HDL = 3,3:1 w/w).t)

the lipoproteins was about the same as in plasma and the fibrinogen concentration was 1 g l^{-1}. We started from a stock solution and then made serial dilutions. The curves for fibrinogen adsorbed to PS are shown in Fig. 1. When you increase the concentration in the case of VLDL you do not see a decrease in the amount of fibrinogen on the surface. There is a little decrease for LDL, but here is an enormous decrease when HDL is present in the mixture.

Thus, fibrinogen adsorption is not effected by VLDL, it is effected a little bit by LDL and it is very much effected by HDL.

The adsorption of HDL to PE after 1 hour as a function of plasma concentration is shown in Fig. 2.

In a simultaneous experiment PE surfaces were exposed to the same plasma solutions. So we did six separate experiments. After 1 hour the 'precoated' surfaces were rinsed and exposed to 1:1 diluted plasma for 1 hour. Thereafter the amounts of adsorbed HDL were determined.

When you begin with plasma dilution of $1:10^5$ you have practically no proteins on the surface. So, you'll find the relative adsorption value of 0.4 which is in accordance with the first experiment (1:1 diluted plasma).

But we know that from diluted plasma (5×10^{-5} to 10^{-3}) adsorption of many proteins takes place. Nevertheless we observe a displacement of all these proteins by HDL. Therefore, there is a preferential adsorption of HDL from plasma to PE.

212

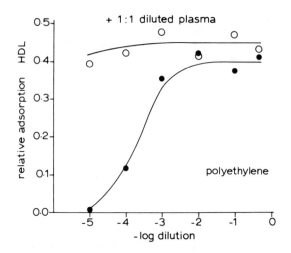

Fig. 2. The preferential adsorption of HDL to polyethylene. Adsorption of HDL from solutions of human plasma to polyethylene. Exposure time 1 hour (●).
In a simultaneous experiment polyethylene surfaces were exposed to the same plasma solutions (plasma concentrations 10^{-5}, 10^{-4}, 10^{-3}, 10^{-2}, 10^{-1} and 5×10^{-1}). After 1 hour the 'precoated' surfaces were exposed to 1:1 diluted plasma for 1 hour. Thereafter the amounts of adsorbed HDL were determined (○).

Lemm: So if it is really of interest to have information within the first 5 sec maybe can we think about how to draw such experiments.

Brash: I think that logically it must be important to know what happens in the first few seconds, because whatever happens after that depends on it. So, by definition, it has to be important and of interest.

May be it's not very good to make an analogy between this situation and cosmology, the big bang etc. Do we need to know what happened in the first fraction of a sec of the Universe to really understand what we have now?

I think that there is a connection between the very early events and the ultimate events. And if we really want to understand what's happening then it is important to look at that.

Like other people here, I'm not aware of any really good techniques which are available to study events at a fraction of a second

Lemm: In applied chemistry there exists experience about precise recording of data concerning the kinetics of high speed reactions.

Brash: Yes, but interfacial events such as we are trying to look at here may require different techniques and I'm not sure that those techniques that you mentioned have been adapted to working on interfacial phenomena.

The technique which I described earlier (Ch. 17) is in principle able to look at things that are happening at a fraction of a second. If things are happening rapidly enough to cause a change in radioactivity at the exit of the cell, then we can measure them. If the events that are happening are not causing a change in that quantity then we can't.

Other events, like changes in conformation, we could not measure by this technique. But we certainly could measure quantities of adsorbed protein at a fraction of a second.

Hoffman: I think your early time measurements are really idealistic. In a sense they're very appealing to a scientist, but they essentially insist upon a system that for all intents and purposes is totally in equilibrium except for the protein coming to the surface.

Take a material that's about to be implanted or contacted with blood. The likelihood that it is in equilibrium at that interface with water, salt and the other components of the system is very low.

Take a surgeon who implants a device. Even if he preclots it, the active preclotting is taking a dry material and mixing it with blood. So, it's a very messy business.

It's very unlikely that you have this equilibrium at the interface with its chemical groups, some of which are polar and like water, some of which are hydrophobic and want to hide from water. Then you bring in your bloody system. It's not a pure protein solution. it's not even diluted plasma; it's blood. So, what happens in the first few instances is in reality a big mess.

What we are talking about is appealing idealistically and scientifically, but I frankly don't think it's worth trying to go after the first few seconds because of this. I think in the end it will be meaningless in the practical world.

Missirlis: Does anybody disagree with that?

Jennissen: I just want to give you some numbers to what was said. I think they are in agreement with what Dr. Hoffman said.

We calculated the half-life for the pseudo first order adsorption rate of phosphorylase (Ch. 8, Ref. 14) and obtained a value of ca. 6 min for the adsorption step. For the desorption step we calculated a half-life of 180 min.

So actually the rates appear to be very slow. There can be fast ones, as Dr. Wolf has said. On a hydrophobic surface as ours however we only observe very slow processes.

Cazenave: I agree with Dr. Hoffman but I think we should be optimistic. We maybe useful to surgeons, because it's a messy business, as you say, but sometimes we have to tell them not to do messy things, in order to improve the thromboresistance of a material.

Brash: I'd just like to say I have to agree with Allan up to a point. One comment you made was that we're not dealing with diluted plasma or blood, we're dealing with blood.

214

But if you're doing any kind of clinical blood contact procedure you are in fact dealing with diluted blood at the very beginning of the interaction. Because normally your prosthesis is filled with saline or some other solution and therefore right at the interface as the blood comes in you probably have a certain volume of diluted blood near the surface.

And it may well be the interaction of that diluted blood with the surface which is the real initial event. So I'd have to disagree on that point.

Hoffman: I accept that. But don't forget, John, a lot of the situation we're talking about concerns temporary implants like catheters. You have problems at the tip and that is very complicated. But not diluted blood interactions at that tip. Not necessarily. . .

Brash: Not necessarily. But it may well be filled with a solution, when you introduce it into the circulation. . .

Wolf: I agree with all the things which are said and I would say that we need the knowledge about the short time events at the surface and I would focus your interest also at the events which are going on if clotting factors are adsorbed and they are activated in a very short time.

So we have to know what happens really in the first few seconds or minutes because then a lot of subsequent events depend on these steps.

Cazenave: I think that if you manipulate very well from the beginning and we have this experience with this fellow implanting shunts, if you improve by avoiding air interface, putting the right buffers or coating the prosthesis with different proteins, you can improve just be doing that, the obstruction time from 2 to 8 or 12 hrs. So I think the first events are important and may be controlled if we understand them.

Jennissen: What is the evidence for these very short term reactions? Do you, Dr. Wolf, have any half-lives for your interactions on the surface? In what time span are your effects happening?

Wolf: The only result is if we measure the change of the surface charge of our particles and we put maybe PS or glass or any particle into a protein solution and we start to measure then after 15 sec when we can make the first measurement the surface has already drastically changed after this time.

So, that means before we can measure all these events , I don't speak about fully coverage of the particles, the processes must be very fast and so I think that is a very important point and we have to look at this fast kinetics.

Jennissen: But you have a charged surface and there could be differences between a charged surface and an uncharged surface in respect to the sorption kinetics.

Wolf: Yes, you know, even if a surface is chemically uncharged if you put the surface into an electrolyte, then the surface will be charged because of adsorption of anything. You find every time negative charged surfaces normally, also if they have not dissociated groups at the surface.

Missirlis: Sometimes in the literature you read that as you bring into contact a material with blood then the first thing that happens is protein adsorption. Can we qualify that? Is it the same as if you put for example an artificial vessel, which means that the bloody business that Prof. Hoffman alluded to it is different than when you put a catheter or some other biomaterial? What is the hard evidence that the first thing that happens is protein adsorption?

Hoffman: Yanni, I would like to follow up with that; it's an interesting question.

I think there is good evidence, just from the molecular size, that the proteins will get to the surface before most of the cells do.

But the fact of the matter is that in some cases we don't get complete monolayer coverage. We're assuming that it's a monolayer. Isn't it possible that along with the protein, in fact within the first minute or so a cell will occasionally contact the surface and see the surface directly? We have one student working with Tom Horbett on this question trying to demonstrate that.

Wolf: I would like to add another aspect. The most experiments which were done with our model systems here, I think are made if the surfaces are put into equilibrium with electrolyte or say liquid so the surface is already in some kind of steady state.

We have made some measurements with colloid chemical method which was introduced by Shukin and that means to measure the adhesiveness of a glass ball put onto a surface and you can measure the detachment force. We have investigated different polymers only to measure this detachment force of a glass ball. We started with a dry surface and what we could find was a drastic change of the surface properties as a function of time.

After, say, 24 hrs, most of polymer surfaces were coming in such kind of equilibrium or, say steady state. We have done also contact angle measurements and also investigated surfaces covered with different proteins and we could find such typical time functions of detachment force.

The conclusion was that if you start with a dry surface, and that is in very often medical cases really the case, then you have additionally to the protein adsorption a process of swelling of a very thin layer of polymer. In other words you have also a change of the surface properties which adsorb the protein.

Hoffman: Did you say detachment force of a glass ball? What is that?

Missirlis: O. K. we have 5 min to end this morning's discussion, so I would like whoever takes the floor now, John, Allan, Adam, to try to summarise the ideas of kinetics of protein adsorption and their usefulness.

Brash: If I'm going to be given this task I'm going to say something first about the question of whether protein adsorption is the first event.

I would say that it's the first *significant* event which takes place in relation to blood compatibility issues. There may well be other things that happen at the surface such as adsorption of small ions or other smaller species in blood. But nobody has looked for those. Maybe somebody should. Maybe somebody will find a connection between those events, if they happen, and blood compatibility.

But it seems to me that at the present state of knowledge, based on the hard data provided in by people like Vroman, using ellipsometry, and even by us 20 years ago, using the old fashioned ATR-IR set up with an ex vivo shunt experiment, that protein adsorption happens very rapidly. And I think it's the first *significant* event that we know of, right now, in relation to blood compatibility.

Summarising this discussion, that's more difficult. What could we say? I think that one thing that came from the discussion was that we need to look in situ if we want to do valid kinetics. And it doesn't matter whether we are concerned about the very initial events or the later events.

Because if we don't do in situ experiments then we may lose Adam's loosely held protein, which may well be a highly significant thing.

So, techniques of this type should be more and more used and that means they should be more and more developed. I'm not sure if we've got a general agreement on that but certainly that's my impression.

We should worry about particles versus flat film. Perhaps we could call that microscopic versus macroscopic surfaces. And as to the 'bottom line' on that I'm not really sure, I think that maybe Adam could add his point of view on that aspect.

Another general point was that if one is going to do valid adsorption kinetics studies one should be concerned about diffusion limitations or transport limitations. Perhaps transport limitations is a better term. One can be fooled in studying adsorption, if you do the experiment for a time which is too short or in a geometry which is inappropriate. You can simply be measuring the rate of diffusion or transport of the protein to the surface. And so if you develop a new experimental protocol you shouldn't go too far without doing that kind of check.

We also discussed the Vroman effect, perhaps ad nauseam, but certainly in depth. I still have my idea on the more general definition of the Vroman effect and I would offer it as such. I don't know that the Dutch group would agree with that necessarily.

Just in parenthesis, I'm not really sure that the data that you showed (Beugeling's) is looking at the Vroman effect. If I go back to your diagram where you showed the concentration domain data for different proteins, the peaks seem to occur exactly at the same concentration in each case. Which is not the Vroman effect. The peaks should occur at different concentrations. So perhaps you're not looking at the Vroman effect and perhaps that's why you don't see it in the time domain. I'm not sure.

What else? The question of events at very short time? There seems to be a majority opinion that we don't really have to look at very short times. I'm in the minority. Personally, I'm going to look at short-term events as closely as I possibly can, because I think that what happens at very short times determines what happens later. And if we think that it's important to find out why things happen then we have to look at the very beginning.

Missirlis: Does anyone wish to add a small comment on that?

Baszkin: I believe that to gain an understanding of protein adsorption it is important to look at these short times.

Hoffman: I don't want to continue on the significance of the short times, but I want to talk about the significance of time in everything we've talked about.

You have the time it takes the protein to get to the surface. Then you have the time of the conformational changes of the protein on the surface. One could be much shorter that the other; the former is usually much shorter. Superimposed on this is the arrival of other proteins with their ability to displace the proteins on the surface.

So, people who are working in this field should be very sensitized to protein displacement and protein conformational changes. In addition the use of antibody labelling to study the displacement and movement of proteins to and from surfaces is potentially artificial because of the change in conformation over a period of time of the protein on the surface and its loss of its recognition characteristics. So, I think the techniques we use also have to be very cautious in terms of these time effects.

19. Protein adsorption and thrombus formation

Missirlis: The questions to be dealt with are protein adsorption and thrombus formation.

The first thing we put down is how the research of protein adsorption promotes the synthesis of better thromboresistant surfaces?

The second related question is how what we have talked so far, that is protein adsorption is a relevant method at all-question mark there-to predict the thrombogenicity of a material in contact with blood?

After coffee break we're dealing with the problem of whether results obtained by these protein adsorption studies are comparable to those of other thrombogenicity tests and we should mention some of these. And then the problem of non-thrombogenic surface and its affinity to proteins comes in all these three aspects.

I'm sure it is very difficult to separate all these subquestions but have them in mind and try to address each on separately if possible.

Brash: One approach to the first question 'did research on protein adsorption promote the synthesis of better thromberesistant materials?' might be to consider a number of materials that we consider to have been developed as a result of *any* research and make a listing of them and then ask the question regarding protein adsorption with respect to each one of those materials. It's a possible way to proceed.

Cazenave: I would like also as a stimulant to put this question the other way around. I think that people who have been synthesizing more thrombo-resistance surfaces have realised that protein adsorption might be very important. I think that they imagined for some time that they could design the best possible polymer, but now that they realise that it's going into a biologic environment they become more humble.

Beugeling: I also wanted to say what Jean-Pierre told us. Chemists know that there is a low protein adsorption onto very hydrophilic surfaces. So they made various hydrophilic surfaces to see whether or not they were thromboresistant.

Y.F. Missirlis and W. Lemm (eds), Modern Aspects of Protein Adsorption on Biomaterials,
219-248.
© 1991 *Kluwer Academic Publishers. Printed in the Netherlands.*

And I think that was another approach in which protein adsorption played a role.

Hoffman: Actually, as I pointed out in my short talk, the most hydrophilic surfaces were not very thromboresistant. But this is what I would like to comment on.

It's like a group of blind men trying to describe an elephant. Each of us has our hands on a different part of this animal. Maybe some parts we're sharing but other parts we're unique in touching.

We do one type of blood compatibility test, you do another type. We reach one conclusion and you reach another. It seems to me the common denominator still is protein adsorption on foreign surfaces. Unfortunately I can only tell you there are very few correlations that I know and believe that make sense between protein adsorption and thrombogenicity.

This has been to me very frankly the big disappointment in this field. You would expect a lot more to have come out of it than has come out. And I would like to hear from my colleagues their experience and their comments to this challenge. I can give some examples where this is a good correlation but again it's the part of the elephant that I'm touching in my testing.

Missirlis: Alan, would you be so kind to list a few such correlations in the blackboard? That is non thrombogenic materials and protein adsorption studies. And then start from there.

Hoffman: To begin with, I have to go back in history.

I'm not familiar with all the different literature in the field but the earlier test was the whole blood clotting test, which is a Lee-White test. It was done in a test tube and there was then a correlation between, fibrinogen adsorption and clotting time. This was using siliconized tubes and glass tubes.

This is during the very beginning of our field where we saw surface composition influence blood thrombogenicity.

This test has long since been discarded because of the air interface and the lack of relevance to a flowing blood-implant situation. A lot of people still use it, but I don't think they believe in it.

Historically, we had a number of different clotting tests. I'm focusing on clotting tests rather than protein adsorption because this is where you have to generalize about what test you're doing to define thrombogenicity or non-thrombogenicity.

Then, we had a number of other in vitro tests that contacted in various cells or various inserts. For example, a paddle test. These tests looked at platelet adhesion and also, I believe, fibrinogen adsorption. Fibrinogen has been the protein that most everybody has studied. And these tests are also discarded today.

So historically, there was the Lyman cell. There was a problem here with the whole blood clotting test. The Lee-White test didn't have flow.

And again, people were looking just at platelets on a surface and fibrinogen adsorption. Then, there was the Lindholm test. Lindholm modified the Lyman test. Again, it was a flow cell test, and again it was clotting time that people looked at, not protein adsorption.

Then there was the in vivo test. One that a lot of people liked and talked about was the vena cava ring test. We used that for many years. This was implanted in a dog in the inferior vena cava and you looked at the amount of thrombus deposited on it as a means for quantitating or qualitating, various surfaces. So, this was a thrombus deposition test.

Brash: Allan, similar to that is the Kusserow ring renal embolus test which was less widespread but was nonetheless fairly important I would say in the early seventies.

Hoffman: Thank you John. That's exactly were I was leading up to, because this was also discarded. The vena cava ring test was discarded as meaningful, because it essentially did not consider the possibility that there was embolization.

The Gott ring test was this little ring of the material implanted in the inferior vena cava if it stayed open or patent for 4 hrs or 2 hrs, then it was allowed to go for 2 weeks and if it was still open this was considered to be a good material.

And hydrogels were very good in these tests so therefore hydrogels were acceptable – that's where it all started with the hydrogels. This was a good material but it was a slippery material and in the renal embolus test it was looking at the kidney infarcts and this was looking for emboli and it turned out that the emboli produced were significant from almost every material. Except as Bob Baier pointed out, in microwave, gas discharged treated glass surfaces, the best material.

So from here I think I can say looking it back at all of these, there was one test that I remember there was a distinct correlation between clotting time in one of these tests here and fibrinogen adsorption. I remember seeing a plot of that, but that was about the only correlation of those early days – John?

Brash: Well, I don't remember that, but maybe my memory fails me. I just don't recall anybody trying to correlate a whole blood clotting time test or any kind of clotting time test with adsorption of any kind.

Platelet adhesion definitely but not clotting. We discussed the example of an albumin layer. We know that it's fairly platelet resistant but we don't know what it would do in terms of a clotting activation. So you may well be right, but you can't remember who it is that did it.

Hoffman: No, I can't remember and it's almost not worth even trying to remember.

Lemm: I mentioned that we found a certain, but not exact correlation between thrombus formation and fibrinogen adsorption in the Nose-Blood-Chamber test.

Cazenave: Logically one would not see why you should have a correlation between clotting time and fibrinogen adsorption. But you do have a correlation between clotting time and factor XII adsorption. HMWK I could understand, but that's not logical, maybe fortiutus. Then you have a correlation with platelet adhesion which will focus the assembly of clotting factors, that makes some sense.

Hoffman: I don't recall any correlation with factor XII adsorption and clotting time. Do you, John?

Cazenave: Well, there is at least one. It's the White-Lee clotting time in a glass tube and in a plastic tube.

Hoffman: With the factor XII?

Cazenave: People who don't have factor XII have a prolonged clotting time in a glass tube too. And if you put back factor XII you shorten the clotting time.

Beugeling: Also thromboplastin time; without adding an activator.

Cazenave: Yes, but thromboplastin will start the coagulation system at another point. It's not the same. It's the extrinsic pathway. XII deficient patients have a prolonged PTT.

Missirlis: I think I got confused whether there are any correlations or not.

Brash: There are no studies. So there can't be any correlations.

Hoffman: I don't know of any correlations of any significance with most of these early tests. People were measuring clotting times and thrombus deposition and later on emboli deposition. There was very little to do with protein adsorption.

What is the best approach in this discussion: to talk about the tests that are used today that are potentially correlatable? Does anybody know of any blood-protein test that they would like to talk about?

Cazenave: Well, going back to the whole blood clotting test, it's not used any more I would say in coagulation, because it's a rather inaccurate test. The end point is very difficult to pick up. It lasts 15 to 30 mins and there is marked variation from one individual to the other depending on the amounts of clotting factors and the amount of antithrombin he has in his blood.

I think it's very difficult to correlate something with this very crude test. It's markedly prolonged only when you have very severe clotting deficiencies. So I'm not surprised that it has been abandoned.

Hoffman: I think a lot of people looked at in vitro platelet deposition and on surfaces that are well characterized and they have also looked at protein adsorption.

Let's start at this point. Who can identify publications or correlations with different materials?

Beugeling: I've already mentioned two materials: that is the pellethane 80A, a PU, which shows a higher fibrinogen adsorption and a higher platelet adhesion than a PU that has been surface-treated with PEO.

Hoffman: That's the latest thesis of yours?

Beugeling: Yes. Another example is the grafting of poly(N-vinyl acetamide), on both SR and PE. These surfaces also show a lower platelet adhesion and a lower protein adsorption than the untreated surfaces (Ch. 4, Ref. 19).

Baszkin: Yes, but there again everything is referred to platelet adhesion.

Hoffman: Most people have looked at in vitro platelet adhesion and there are many people who don't feel this is a valid or relevant test to what's happening in vivo. The literature is filled with many correlations that show as fibrinogen adsorption goes up, platelet adhesion also goes up. And vice versa, the opposite is true, that as fibrinogen adsorption decreases platelet adhesion seems to be less, in vitro.

Baszkin: This is in agreement with Lyman's hypothesis and with the results presented here by Cazenave.

Hoffman: Fibrinogen is one of the simplest ones to correlate. Are there any other proteins that seem to correlate? Albumin,

Baszkin: Also collagen, according to the results of Cazenave.

Brash: Mustard published 20 years ago on IgG as a platelet-active protein (M.A. Packham et al., *J. Lab. Clin. Med.* 73, 686, 1969). He found that a layer of fibrinogen on glass was platelet active. A layer of γ-globulin on glass was also active, not only to make platelets stick but to make them release as well.

And in fact in their hands IgG was more reactive than fibrinogen. But that has never been followed up by anybody. IgG has been neglected as a possibly sticky platelet protein.

Hoffman: I could also mention fibronectin, which is well-known as the adsorption of fibronectin increases platelet adhesion will also increase.

Beugeling: I believe that's only for precoating a material with fibronectin or vWF, which enhances platelet adhesion.

Hoffman: In a sense it doesn't matter for precoating. We're looking for any kind of correlation with protein adsorption.

Beugeling: vWF shows the highest platelet adhesion.

Baquey: I would like to say that when you want to make a correlation between the in vitro and ex vivo or in vivo results you must consider the time of exposure.

And in fact we can't find a good correlation between the in vitro experiments and the first minute of exposure in vivo; and if duration of exposure is increased, thrombosis is happening. So it's difficult.

If you look only at the whole result you will have a material which seems to be thrombogenic despite good results dealing with protein adsorption. And if you look at initial events you don't have neither platelet adhesion in vivo, nor fibrinogen adsorption or accumulation like in vitro.

So, you have the same results in vitro, and in vivo during the first minute, but afterwards other factors are involved, such as haemodynamic factors and the chemistry of the surface alone is no more efficient to control what is happening at the interface.

Cazenave: I think one also has to take in consideration, if you're dealing with vWF, that the shear rate is very important. It will be an adhesive protein at high wall shear rate and not at low wall shear rate. I think that's very critical. So there are different types of adhesion of platelets depending on the flow conditions. In our experience fibronectin platelets will stick to it, but it's not better or worse than fibrinogen.

Baszkin: One simple question to address to people who are more close to this problem. Is platelet adhesion responsible of clotting and if it is to which extent? And if it is not why we look at platelet adhesion?

Cazenave: Well, platelets are not responsible for clotting, but when platelets are activated they will change their membrane orientation of ionic phospholipids and expose anionic phospholipids (phosphatidylserine) from the inner leaflet on the external leaflet and make possible the assembly of clotting factors.

Coagulation is a surface controlled reaction and one of the difficulties is that most of our tests of coagulation in vitro are in the fluid phase, whereas coagulation in vivo is always at a surface.

So, platelets are important, in a way, in coagulation, because they enable the assembly of coagulation factors and will multiply the formation of prothrombin as complex 300000 times.

Baszkin: So, if I understand well, platelet deposition on the surface will be an indicator of some unwanted effects. Will the coagulation phenomenon occur?

Cazenave: May occur. It will localize thrombin generation and fibrin formation.

Beugeling: May I add the following to the remarks of Cazevane. There is a mutual interaction between platelet adhesion/aggregation and surface activation of blood coagulation. Because during surface activation of blood coagulation thrombin is formed and thrombin will activate the platelets. On the surface of platelets, which are adhering and aggregating, blood coagulation takes place. Thus, there is a mutual interaction in which both processes are stimulated.

Baquey: You can see in the first column (Ch. 1, Table 2) the comparison of the kinetics of platelet adhesion for medical grade PE and PE which has been modified according to the heparin-like procedure. That's an ex vivo experiment; in vitro the modified PE has been shown to have less affinity for platelets than the non-modified PE.

But as time is increasing you can see that the kinetics are progressively larger and larger. And after more than 1 hr of circulation the kinetics is becoming equivalent to the initial kinetics of adhesion on the non-modified PE.

So if you compare the two materials in a short time scale you will find a good correlation, for the platelets, between the in vitro an ex vivo experiment. If you look at the fibrinogen adsorption you can see that for near 15 min you have no platelet adhesion; here you have a decreasing slope for the fibrinogen which means that there is not to much fibrinogen accumulation; and at a given time, I don't know why, there is a platelet accumulation, fibrinogen is still not adsorbing and later there is a little more adsorption or accumulation of fibrinogen concomitant with an embolization process.

And you can see that there is a variation of the volume because there is an increase of red cells count, which is the index of the volume which is taken into account.

So, as a conclusion, it's a question of time scale. You can only compare the initial events. It is my main opinion.

Jennissen: What are the initial events you're talking about? What do you mean with initial events, e. g. in relation to the time the platelets adsorb?

Baquey: By initial events I mean the ones which take place during the first 30 sec or first minute. In this case I think that the results are quite comparable in vitro and ex vivo during several minutes. But after several minutes it's no more the same thing.

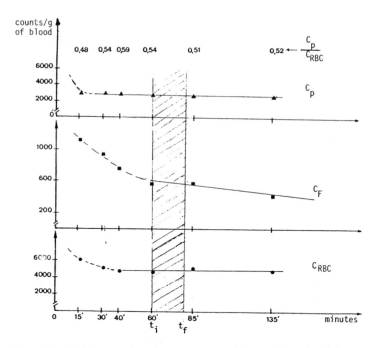

Fig. 1. In vivo experiment. Changes over time in platelets, red blood cells and fibrinogen concentrations (C_P, C_{RBC}, C_F). Each of these concentrations is expressed by the blood volumic radioactivity related to the corresponding radiotracer and measured for blood samples coming from the whole circulation; t_i and t_f limit a period of time during which blood is allowed to flow through the extracorporeal shunt or by-pass settled by the surgeon.

Jennissen: Yes, but what events do you mean? And what is the evidence for them?

Baquey: Obviously they cannot be at all the same. Because in vitro a blood cell suspension is used in the system which has been well described by Jean-Pierre Cazenave and Tom Beugeling, and in vivo it's whole blood. So the things cannot be completely related. In the first in vitro system there is not too much fibrinogen available.

Jennissen: Yes but in your system you did not show us any initial events which might be the cause for the platelet deposition.

Baquey: I don't know the cause for the platelet deposition. I think it's mainly haemodynamic factors.

Jennissen: It could just as well be another protein being adsorbed and not fibrinogen.

Baquey: No, I was not saying fibrinogen was responsible because you can see (Fig. 1) that the platelet accumulation is increasing steeply before the slope of the fibrinogen curve is changing. So there is no relationship between fibrinogen accumulation, probably, and the platelet deposition.

Brash: I want to make two points. I think that to introduce the complication in vitro versus in vivo makes things difficult. And maybe this discussion would be easier if we tried to think about in vitro correlations between some indication of thrombosis (either clotting or platelet thrombosis) and protein adsorption. Then we could try the same for in vivo correlations between in vivo protein adsorption and thrombus formation.

Missirlis: I think that's the way to go, I think and could you please continue on this. I mean in vitro correlations of some protein adsorption and...

Brash: I think the best work in this area, certainly the most extensive, has been done by Stuart Cooper. It depends whether one considers in vivo and ex vivo the same or different. Maybe we should mention his experiments in relation to in vivo.

We have already discussed correlations between platelet adhesion in vitro and precoating of surfaces with different proteins. It appears now that there also have been results where people have observed fibrinogen adsorption and platelet adhesion simultaneously, and shown that these things are correlated. That's what you've done, I think, isn't it Jean-Pierre?

That was lacking for a while it seems to me. All the work had been done with *precoated* surfaces but it had not been shown that fibrinogen adsorption was going on in a way that could be correlated to simultaneous platelet adhesion or platelet interaction with the surface.

So that seems to be something which has come along in the last little while. But I should leave it to other people to fill in the gaps regarding the in vitro aspect.

Baquey: We did not do it simultaneously. We measured protein adsorption on flat surfaces and the platelet adhesion in the Cazenave system. So it was not simultaneously but the same materials were involved.

Hoffman: Can I put a few slides up? What I prefer to talk about is our own experiments, which are very definite correlations that we have found with in vitro and in vivo blood responses and protein adsorption. This is just one system.

Missirlis: While you're looking for your slides when we talk about these correlations do we mean that the surfaces we're talking about are very well characterised? Has any work been done on these IUPAC materials?

Brash: The work is going on and will probably be reported upon within the next year. But it's on going.

Lemm: We can just switch to the first question. Did really all this research promote the development of better thromboresistant surfaces up to now? My opinion is no.

Hoffman: I agree with you.

Brash: I don't agree. I think that protein adsorption (being of course an advocate of this field) has given important indications. If it hasn't defined better surfaces then it has given indications of directions to go in.

Lemm: But for example we could modify surfaces in different ways that we selectively adsorb proteins which are good for thromboresistant surfaces. But it did not happen in the past.

Brash: With respect to some of the surfaces which are being developed at the present moment, one could argue that the rationale for them, is based on knowledge of how proteins may interact with them.

Missirlis: I think what you just said if we have some evidence for that. Is there for example somebody who produces a new materials, who says that he took in consideration this and that study on protein adsorption?

Brash: All the surfaces which people are trying to develop based on 'zero protein adsorption', the non-fouling surfaces, are of course being developed because people think they won't adsorb proteins.
  Whether that is valid or not is of course still debatable but I think that's got to be an example.
  I would say that the heparin-like surfaces of Marcel and Jacqueline Jozefowicz (M. Jozefowicz and J. Jozefowicz, *Pure Appl. Chem.*, 56, 1335, 1984), which I think are the surfaces you just showed, are based on the rationale that they'll selectively adsorb antithrombin III. At least they'll adsorb a relatively large amount of antithrombin III. That has to be the case if indeed the mechanism of heparin-like activity is correct. So there is a second example.
  I think that Bob Eberhart's surfaces, where he grafts long-chain alkyl groups to polyurethanes to promote selection of albumin from the blood is another example (M.S. Munro et al., *ASAIO Journal*, 6, 65, 1983).
  There are quite a lot of people doing this kind of development based on this rationale, but there is a question whether the surfaces have been sufficiently developed to decide if they are thromboresistant or if the approach really is a good one. The development hasn't gone far enough at this point. I would argue that the rationale for all those surfaces is based on ideas that come from the study of protein adsorption.

There are probably other examples besides the three that I've just given.

Hoffman: The hindsight is we started a process. This was a project where we were going inside the vascular grafts with a fluorocarbon coating that we could compare Dacron having a Teflon-like coating with Gortex. We were going to compare different Dacrons of different porosity and different, essentially net versus weaved versus Gortex.

So we treated the inside of this graft. The gas goes through, you get a gas discharge here, which when you move this along it deposits a fluorocarbon coating on the inside of the graft. So then we put it onto a baboon shunt.

This is Steve Hansen's data and I showed you this also. These are the numbers of platelets deposited per unit length. This is an ex vivo shunt in a baboon. This is the untreated Dacron here and this is the treated Dacron and as a function of time. I agree with Charles Baquey, that you can do in vivo studies, but usually they're very short term, and whether they are relevant towards the implants is already a major question. But it's the best we can do right now. So you see a major difference in platelet adhesion.

And you ask yourself, how does this relate to protein adsorption? It's a backwards type of approach.

The next slide emphasises that there is a lot of thrombus deposited on the untreated Dacron and very little which is visible on this treated Dacron.

So now you ask, what's inside there? A lot of platelets. We've seen the In-labelled count in the previous slide. Are you making thrombus here and essentially embolizing the thrombus off? If you are, this is very dangerous, so then you have to check on embolization.

Here are our data an emboli volume from a laser scattering system, a function again of time. In this period of time, 1-2 hrs. You see that the untreated Dacron has a significantly greater amount of emboli produced than the treated Dacron, emphasizing that a clear surface gets around the problem with the vena cava ring.

Now we've got in vivo data or ex vivo data. These data were reproduced ex vivo, but these happen to be an in vitro set of studies. Now the question is, how does this relate to these initial events of proteins in the first few minutes or seconds?

These are explants in the ex vivo shunt up to 1 week. So we went even further. Instead of 1 hr or 2 hrs we went up to 1 week. We found here that the grafts that were treated, which are the dark bars, remained patent or open, as a function of time after implantation in the shunt much more efficiently than the untreated, which were closing off very rapidly. And this is for 8 different baboons in each case.

So, we have a lot of data that we've accumulated on a surface and we have yet to characterize it in terms of protein adsorption which is the main question. So, we went back and began to look harder at protein adsorption.

The first thing people do in the field of protein adsorption is they look at fibrinogen. Fibrinogen is the bad protein. That's the first thing, of many others as Jean-Pierre has mentioned that are important proteins in the coagulation cascade. This is the fibrinogen adsorption to the untreated.

Now you step back. We were looking at a fabric with a flow regime that had a certain characteristic. Now we're looking at a piece of film sitting in whole blood or blood plasma (actually it was diluted, perhaps 80%) in a beaker.

So, now you've gone away from the flow situation. You're looking at a somewhat idealized protein adsorption study in vitro, the amount of adsorbed protein versus time. Here in the first few minutes you see a large amount. You don't see that with the tetrafluoroethylene treated flat film of the same composition as the fabric. So, this is the first indication.

There is a difference in that protein adsorption. So, we went further. We want to know what is the character of the protein on the surface, even after perhaps 2 hrs. And after several hours we looked at how easily that fibrinogen could be washed off the surface, whether it was untreated or treated.

Here you can see if you leave it on for 30 sec or for 2 hrs that the percent which is washed off or ellutable is essentially decreasing with time as the time on the surface increases. It's also significantly lower after the treatment.

So, we say essentially our treatment has produced a surface that adsorbs less fibrinogen which is bound to that surface. In fact, it's very difficult to elute off. So it's a different kind of fibrinogen than the untreated surface.

The idea that fibrinogen which sticks on a surface is not a good kind of fibrinogen. So there must be something else that's happening.

And we've looked at albumin and we found these results which I also indicated earlier, that the amount of albumin adsorbed on the surface of the treated PE teraphthalate is about the same as the untreated. But then when we try to wash it off. The amount which remains on the surface of the treated is perhaps 90-95% of the adsorbed amount where as in the untreated surface it's under 50%.

So, here is a correlation with the thrombogenicity test in vivo and in vitro. It's one of the few that I know of, but we still don't understand it. We don't understand what it means if the albumin is difficult to wash off the surface. We don't understand what it means if fibrinogen also sticks on that surface. And that's supposedly not to good.

The conclusion I would reach, is even though this is a direct correlation of a thrombogenic behaviour with protein adsorption, it's in its infancy. It's the very beginning of what we don't really understand. I don't know of very many other correlations. And I haven't heard them here yet either.

So, I want to make sure the people who have not worked in this field for a long time, walk away from this meeting with this conclusion: Be careful of any generalities of protein adsorption correlations with blood thrombogenicity.

Baszkin: I'm happy that you have shown this because it's a very good approach to the problem. I'm very satisfied of this demonstration and I think it's a good way to do the things. To observe a phenomenon, to go backwards to see what happens and to try to explain.

Cazenave: I would like to go back to what you said. The correlation game, I think, is a very difficult one. In another aspect people have been for years and years trying to correlate what they call platelet arterial aggregability and arterial disease. And there is no correlation or easy correlation between in vitro test and in vivo tests. So, I'm not surprised.

There is another example, which I think is very similar. It is not with polymers, but when you damage in an animal the aorta by a balloon catheter you see over time a modification of the vessel wall so platelets will not adhere to it. With time this is constantly modified.

I'm sure that the surface of a material will also be modified with time. Plasmin may be generated, fibrinogen may be degraded to fibrin which is not reactive to platelets if there is no thrombin around, leucocytes may liberate elastase which will degrade some key proteins, so I think the phenomena are very complex and there is a lot of work to do, which will be very interesting, but it's not simple.

Jennissen: You showed that between the treated and the untreated surface you have about a factor of two difference in fibrinogen adsorption, but your difference in platelet adsorption was at least 10 fold. So, where is the correlation?

Hoffman: They're in the same direction. I mean we're satisfied with small results. This is no complex. And what I meant when I said originally about blind men describing an elephant: this is our test, this particular shunt. If you happen to run a vena cava ring test, because that was your favored test, you would perhaps get the opposite results and conclusions. So, that's why I say:be careful. What test you use to reach conclusions about thrombogenicity to begin with is already put you in a prejudicial situation.

Cazenave: Again, what you inject into the animal is labelled fibrinogen. What you detect on the surface with a γ-camera is ^{125}I. But there is a clotting mechanism, so you don't know in exactly what form is the radioactivity on the surface.

I think this is quite important. It's a very difficult point to investigate, it might be not fibrinogen any more. And some surfaces may have the possibility to react differently with the clotting mechanism in the body. It might be in the form of some degraded fibrinogen or some form of fibrin.

Baquey: I quite agree with that what Jean-Pierre is saying. In fact we don't know if the fibrinogen is still fibrinogen. And maybe at the beginning it's still fibrinogen but as the time is elapsing there is no more fibrinogen and maybe it could be fibrin. And now we have the opportunity to look at the state of the material which is deposited using monoclonal antibodies which are able to make the difference, which react with fibrin but not with fibrinogen. So, I think we could have the answer in a few time.

Brash: I think with respect to Jean-Pierre's comment there are two things: one is that it's good to be skeptical and one certainly should be skeptical and think of all the possible things that can go wrong as well as the things that can go right. But on the other hand I think has to have a certain degree of faith in what one does, i. e. strike a balance.

And the second point, with respect to fibrinogen, is that in fact 'normal' fibrinogen is not a well defined, precisely described molecule. There is the work by Mosesson, going back 10 years, in which he showed that fibrinogen as it normally exists in the circulation is not one single molecule. It's a 'distribution' of molecules, consisting of fragments of different kinds (D.K. Galanakis et al., *J. Lab. Clin. Med.* 92, 376, 1978). The composition is variable over time.

Cazenave: But it's not fibrin. When I look skeptical, I'm not negative, I mean this is one of the key issue. The fate of fibrinogen is to be attacked by thrombin and plasmin. So it's very difficult to know in what form it is. I think this is very relevant to thrombogenicity and one will have to look into that in the coming years. And it's a very difficult task.

Brash: I agree. Especially in vivo types of experiment where people measure fibrinogen adsorption and they don't know if it's fibrinogen or fibrin. That's a key point I would agree.

Cazenave: In vitro it's easier because we have isolated system or single proteins, well, it's less difficult.

Beugeling: I want to make an additional remark on what Prof. Baquey has said. The monoclonal antibodies are very risky to use because it is possible that the antigenic determinant of the adsorbed protein is directed towards the surface. And then you see nothing but still the protein is there. So we must be careful to use monoclonal antibodies.

Hoffman: I would like to make a philosophical comment in terms of advising young people in this field and of ourselves who are in the field for many years.

I think one thing we need is a very good biochemist, or blood protein chemist to work with. Everybody has to have that on his team or you're going to be fooling yourself, probably doing potentially meaningless experiments.

Secondly, I think if you look at the number of in vitro tests that we can run, I think, there is no such thing as one simple in vitro test to attempt to correlate with an in vivo test that you happen to be working with. I think you need a battery of tests. You need many different tests. Obviously you need a protein adsorption test that is as close as possible to the end condition of your in vivo test. If you have a flow system, try to simulate a flow system with plasma and labelled protein in the plasma to come as close as possible to the flow patterns of your in vivo As I say work with a biochemist, protein biochemist, who knows, hematologist, who knows these proteins well, like Jean Pierre and others in this room.

Furthermore, do many different tests. Do as many as you can think of. Do your potentially competition tests of two proteins, do the labelled protein in plasma do whole blood test. Even do, you know, I hesitate to recommend this, even do some platelet adhesion tests in vitro, if you find them meaningful, but again, as close as possible to the in vivo situation. And maybe do several in vivo tests.

In the end, if you rely on only a few tests, and in your own intuition and some literature recommendations you may and up in a dead end. So, just be careful.

Brash: I would like us to look at what's written on the paper that we're discussing right now and there are two points. One says: does research on protein adsorption promote synthesis of better thromboresistant materials?

I would again ask the question: did *any* research promote such synthesis? Can anyone really think of any research on blood material interactions, not just protein adsorption, which promoted the synthesis of better thromboresistant materials?

And then a second question is: Is protein adsorption a relevant method at all to predict the thrombogenicity of a material in contact with blood?

Then I would ask the same sort of question with respect to that. Most people here seem to be saying that the answer is 'no', but what else predicts the thrombogenicity of a material in contact with blood?

So, I am asking the question more generally if you want. What else is there besides protein adsorption to use as a predictive tool?

Hoffman: I can give one quick example of how material research has led to better materials and that's heparinization of surfaces. That's not a new polymer, but it's a new combination polymer.

Brash: Allan, that was one of the very first ideas, back in 1960-something, that anybody ever had and it is based on the fact that heparin is a known anticoagulant.

Hoffman: But it works.

Brash: Yes, but research on blood-material interactions hadn't even begun at that point in my opinion. So, with respect, I would say that that concept did not emerge from the intensive research which has gone on during the last 20 years on blood-materials interactions.

Baquey: I think that protein adsorption studies could promote indirectly the synthesis of better thromboresistant surfaces, as far as several teams are interested for the synthesis of materials able to be covered with endothelial cells. And you can achieve such material if you are ready to control how endothelial cells adhere preferentially and grow on such surfaces.

234

So as far as you know which proteins are important to promote cell adhesion in vitro, you may succeed in achieving such materials. Study of protein adsorption on a material could be a way to prepare a surface in order to achieve such a material.

Baszkin: I believe it is an important statement. I would like this to be recorded.

Beugeling: I want to go back to the question whether or not these protein adsorption studies have led to better thromboresistance materials.

We began with the Lyman hypothesis and it was found or it was stated that albumin adsorption on PU should lead to a thromboresistant material. And that lead to the synthesis of many PUs and today we have some PUs which are more or less blood compatible. Like the Pellethane 80a. So, in my opinion, protein adsorption studies have led to better materials.

Hoffman: That's a good example. Don't you think?

Brash: I don't know what the evidence is about Pellethane being especially blood compatible.

Hoffman: Survives as a pace maker lead.

Lemm: The problem with Pellethane and with Biomer is that it starts to degrade after 5 or 6 months.

Brash: What happened to the artificial hearts that were implanted in humans?

Lemm: Pace-makers?

Brash: I don't mean the pace-makers. I mean the hearts, the Utah hearts, which were implanted in people like Barney Clark, and others, back about 6 years or so. What was the problem with those?

Hoffman: Apparently the problem was the design of the inlet and outlet sections with various connectors put together with rough edges and surfaces over which the blood flowed and created thrombus. Thrombus deposited; that came off as emboli and ended up stroking the patients. That was the major problem.

But that wasn't necessarily a material problem. I think it was a design problem. And that's one of the problems in doing in vivo studies, the design of your particular system has to take into account hemodynamics.

Brash: Is that speculation, Alan or is that...

Hoffman: That's speculation by people like De Vries and others in the field. How can you prove where the embolus comes from?

Brash: I think there were footprints left on the surface of the PU, which...

Hoffman: That's been pretty clean.

Brash: I think there were indications...

Hoffman: From calf studies or human studies?

Brash: From the explanted hearts from the human patients.

Hoffman: I have slides from the inlet and outlet...

Brash: These were pretty clean surfaces, but the fact remains the end result was a stroke.

Hoffman: The tendency today is to mold it in one piece to avoid these irregularities in the inlet and outlet sections. Now, that's where they're going.

Cazenave: Again, this will be philosophy. We want to design the best surface or the best biocompatible material, which I think is fine and a nice objective. But we need to increase, and this has been done of over the years, the knowledge of the mechanisms which are involved.
It might be that there is not one perfect material, but a number of steps one has to put together knowing the material, the protein adsorption, the biology and the drug manipulation in order to be successful. It might be that the final solution will be rather complex, well, multifactorial. This is what has been done with implanting cells on surfaces or using protein adsorption or drugs.

Missirlis: Can one correlate or say something about the pseudo-endothelization of surfaces? Is that another line of approach to biocompatibility which is relevant at all to our discussion or should we leave it totally out?

Brash: I would say it's an approach which is very 'hot' at the moment, i. e. endothelialization, seeding etc. But I think the relevance to protein adsorption maybe somewhat obscure unless you mean how to fix the endothelial cells onto the substrate. It's not prevention of thrombogenesis by a protein layer.

Hoffman: This is the way we rationalize. In our research group, we are also attempting to develop surfaces onto which endothelial cells can be seeded and will grow.
How do you do this? You take a surface that is not very good and you try to modify it's chemistry so that now adsorbs (preferentially or strongly), fibronectin. That's the model protein for sticking cells.
So we have gas discharge treatments, for example of methanol or acetone, which show enhanced adsorption of fibronectin. Then we seed endothelial cells

236

on these surfaces with the hope that they will spread and grow more easily than on a surface before that treatment.

This is the approach we take. Again, it's with the rationalization that a certain protein is important for this particular process.

Brash: With respect to Allan's comment about trying to promote fibronectin adsorption so that endothelial cells will stick, is anybody trying to use gelatin by any chance as a ligand for fibronectin?

Cazenave: Yes, that was done in the early days. The big advance in growing endothelial cells came when people realised that you had to coat with an adhesive protein. So they tried to coat with gelatin which adsorbs fibronectin from the serum in the culture medium. That does some good, but it's not the best thing. It's better to coat the tissue culture dish with fibronectin.

Baquey: I have another point to add about that. Mittermayer in Aachen (Germany) showed that when fibronectin was covalently bound to the surface, and as far as I can remember that was a PU, when the fibronectin is covalently bound the results are better than when the fibronectin is only adsorbed. So the way the adhesive protein is combined to the surface is important to the following.

Cazenave: Fibronectin is good for endothelial cells but there are other cells which need other adhesive proteins, like laminin for example. So it's not true for any kind of cell you grow.

Missirlis: How long these endothelial cells are good for the purpose they are put in there. How long they survive?

Cazenave: In tissue culture?

Missirlis: No, in vivo. Are there experiments in vivo?

Cazenave: A few experiments. In humans, endothelial cells don't grow from the edge onto, let's say, Dacron prostheses or Gortex, but once you have a layer of cells. It will produce the right extracellular matrix and then there will be a replacement by addressing cells. So the problem is to go through the few weeks necessary for repopulation and eventually turnover of cells.

Beugeling: Fig. 2 shows the results of experiments of Poot and Dekker (Ch. 4, Ref. 14). The experiments were carried out with the Cazenave flow system (Ch. 4, Fig. 1). The PE tubes were first coated with crude fibronectin and thereafter different amounts of endothelial cells were seeded upon the surface, so that the density of seeded cells on the surface was different.

Many platelets stick to the fibronectin surface. When there are more cells on

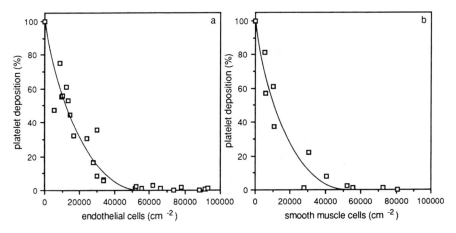

Fig. 2. Effect of human endothelial cells (a) and human smooth muscle cells (b) coverage on platelet deposition, expressed as percentage of deposition in capillaries which were only precoated with partially purified fibronectin.
Perfusions were carried out for 5 min at a shear rate of 300 s⁻¹.
Perfusates contained human platelets, red cells, fibrinogen, Ca^{2+} and Mg^{2+}.

the surface, platelet adhesion is decreasing until the surface is completely covered with endothelial cells and then you do not see platelet adhesion any more.

Our aim was to investigate the role of endothelial cells in this case. Were they producing prostacyclin so that platelet adhesion was prevented? Therefore we did the same experiment with smooth muscle cells (Fig. 2b) and we saw the same result. We didn't expect that.

In another experiment (Fig. 3) platelets were activated with calcium ionophore. Isolated platelets were first labelled and activated by calcium ionophore 3 min before the experiment.

Many platelets stick to the non-covered (fibronectin) surface. There is a rather low adhesion of platelets on smooth muscle cells. But, on the endothelial cells you see a very low platelet adhesion. So we think that this is a real effect; endothelial cells are better than smooth muscle cells in preventing platelet adhesion.

Baszkin: To what extent platelet adhesion is an indicator for screening polymers as far as their blood compatibility is concerned?

Brash: Platelets – Adam, what's your problem with that?

Baszkin: My problem is that it is not very clear. I'm attending many meetings and I cannot understand to which extend platelet adhesion is an indicator of thrombosis.

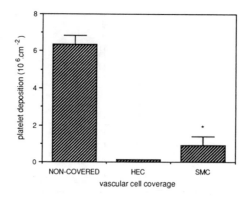

Fig. 3. Deposition of activated platelets to capillaries covered with human endothelial cells
(HEC) and smooth muscle cells (SMC).
'Non covered' refers to capillaries which were only precoated with partially purified fibronectin.
Ca^{2+} ionophore was added to the perfusates 3 min before perfusions, which were carried out
for 5 min at a shear rate of 300 s^{-1}.

Cazenave: Yes, well, there is a good indication which is a clinical indication.
When you have obstruction of implants in humans, one way to prevent
thrombosis without removing the graft is to give antiplatelet drugs, specific
antiplatelet drugs, or monoclonal antibodies which will prevent platelet-platelet
interactions. If you do that the graft remains permeable.

One also has to say that if you inhibit thrombin, it works too. So, I think
these are the two key components. One may expect that if you prevent platelet
deposition or coagulation activation you will prevent thrombosis.

This is obvious, because thrombosis is the activation of platelets the
generation of thrombin and the production of fibrin. Of course it's also the
autologous deposition of endothelial cells on the graft. You have to recall that
they are an accessory cell in immune reaction, so you cannot transplant
endothelial cells unless you have the same histocompatibility antigen match,
which is very difficult. If you do cross individual transplantation you have a
graft versus host reaction, so you are in even more trouble.

Lemm: We all know that the non-thrombogenic surface does not exist till now!
But two approaches to design a non-thrombogenic surface are imagin-
able:either the proteinophobic surface, which does not adsorb any protein at all
or the proteinophilic surface which adsorbs the right proteins whatever they
are?

Which is the most promising strategy?

Hoffman: The major effort in our group right now, to try to decide this
question.

I have a large grant from the NIH to develop non-fouling surfaces, I have a
major effort on these fluorocarbon-deposited systems which tend to bind

albumin, we don't know if this is a passivating albumin or not, and we are trying to compare these two and other extensions of them. We do not know, the answer to that yet, but it's a fundamental question and at the top of my head of my research group right now.

Cazenave: One of the difficulties when you think about what is, say nature's best haemocompatible surface, which is the vascular side of the endothelial cell, well, the difference with artificial surfaces is that living cells have a number of systems which will take care of any trace deposition or activation of the system.

Endothelial cells can release substances, which will prevent platelet activation or deaggregate platelets, will generate ways to cope with thrombin (there is antithrombin III). There is heparan sulphate on the cell surface, there is thrombomodulin, or the cell will release fibrinolytic enzymes.

So this is the difficulty we want to design a perfect surface. It might be best if it won't adsorb anything, so that's a kind of negative approach which could be positive. But it's not a living surface, so there is no inbuilt process which will take care of traces of damage.

Beugeling: We are following two directions at the moment.

In the first one PEO is grafted on surfaces to make them a little bit hydrogel-like, and we hope that these materials adsorb less protein and will show less platelet adhesion, and surface activation of blood coagulation.

In the second one we are trying to make such materials that endothelial cells will stick on the surface when you seed them, and that they will stay on it. In that respect we are working on small diameter vascular grafts.

In this case you have to use autologous endothelial cells. In fact you have to isolate them from the patient, e. g. from capillaries of the fat tissue, culture them during 14 days and then you have probably enough cells to seed on the surface.

Brash: I'd like to just make two points. I'd like to make a comment on Jean Pierre's comment and then answer your question regarding zero protein or the right protein.

I think that you're implying that may be we should be trying to go towards something that resembles an endothelial surface to get the ideal non-thrombogenic, thromboresistant surface, which probably is correct but will be extremely difficult to do unless we can find a way simply to line the surface with endothelial cells themselves.

On the question of whether we need the right protein, zero proteins etc., I think there is some evidence in the literature that the normal vascular wall itself does not bind proteins to any great extent from the blood.

Mark Hatton for example in the thrombosis group at McMaster has done in vitro studies using excised vessels which admittedly may not be completely intact. But his results, I think, suggest there is very little binding of proteins from the blood itself.

Cazenave: It binds proteins for which he has receptor and Marc Hatton and I have shown that within it binds thrombin for example (Ch. 6, Ref. 16).

Brash: Yes, that's right. Binding occurs when there are receptors. But in general, the endothelium doesn't bind other proteins. So we get back to your point on having to have ligands on a surface to bind cells via receptors.

In response to the initial question which was: 'do you want no proteins at all or the right proteins', I think it's not clear that it's one or the other. I would say that zero proteins is something that people are working hard on at the moment and I would support the idea as long as zero protein equals zero cells as well, which is the point Allan Hoffman made yesterday, and I guess the answer is not clear on that.

Is a surface which will not bind any proteins also going to be a surface which will not bind any cells, in particular platelets with respect to thrombogenesis?

So, we have to wait for the answer to that I believe, but I think that it's worth pursuing. Because if there are no proteins then there won't be any specific ligands for cell receptors to bind to. But they may bind anyway.

The 'right' proteins is the other approach and I'm a proponent of this, but again with one reservation, and that is as long as there are no other proteins besides the right protein. And I think that the kind of surface that one needs to do that is the kind of surface that the chromatography people develop. For example, in affinity chromatography, you have sepharose-type materials which are designed exactly so that they don't bind proteins, so that you'll get a size exclusion mechanism in the chromatography procedure. And then what is done for affinity chromatography is simply to attach specific ligands to those sepharose gels.

So, it seems to me that an ideal surface, if you know the right protein, is a 'sea' of inert gel of some kind with specific ligands to interact with the protein that you think you want.

Baquey: If we return in this proposition one can prepare a good blood compatible surface with the right properties, that means able to catch the right protein, then you need the right ligands fixed on the surface.

But the question is to get the right distribution of the ligands. As I said before, you don't need a statistical but rather a well-organised distribution of these ligands.

Because you are working with an open system (flowing blood) and if bad events are to happen they will bring all the system to an unstable situation. So, you must prepare a surface able to catch the right protein, but all the surface must feature this good performance. And that's difficult. It's a question of chemical engineering or biological engineering to get the surface with not any hole in the coating.

Brash: If I may come back again, I think it might be interesting to discuss what people think of the right proteins, to get blood compatible surfaces. Is it one protein or should we design a surface which will attract two proteins. If so, what are the proteins and what should be the ratio?

Hoffman: I think after these discussions we can take a crack of that. I was talking with Jean-Pierre during the break and he said that thrombin is the key protein that you don't want around. You don't want thrombus deposited on a surface.

And I said well, what's the best way to get thrombin out of and he said hirudin binds thrombin fairly, but it binds, most likely in a one step process and then you've killed off the hirudin binding site.

So this just leads me into one comment. And that is the right protein that you want on the surface if it's going to create, if it's going to bind something it should bind it reversibly and so it changes it to be in an inactive form. And protein like this would be for example an enzyme, which continuously turns over and turns over and turns over.

Antithrombin III is another one that has been bound by, for example, heparin, which when it binds thrombin it changes ii into a configuration which is no longer active, but then it's active again as a catalyst. So this combination, the catalytic event stimulated by heparin, leaves the surface still active for further action.

Fibrinolysin is another one. This is an enzyme which dissolves fibrin if it's deposited on the surface.

So these may be the kinds of proteins, maybe not those particularly, but enzymes and other proteins that can have a longer life than just a one-step binding process.

And of course albumin, everybody's favourite.

Beugeling: I want to add HDL to these compounds. Perhaps LDL too, because PE, precoated with these lipoproteins showed no platelet adhesion in the Cazenave system.

Missirlis: Would that be the same needs for say a catheter and a long term device?

Brash: For a short-term catheter implantation you can heparinize the patient and get away with it by doing that. The same as you can with hemodialysis and heart-lung by-pass and other short term procedures. But if you want to do a permanent implantation, clearly you have a different kind of problem.

Hoffman: Something else that Jean-Pierre said was that, we may want to develop, or in the end we may have to develop, the right drug regime to go with an implant. In other words, no implant is perfect. Every implant is going to cause or initiate some kind of thrombosis on foreign response that will create some kind of deposit of thrombin, rather of thrombus. So, the combination of certain drugs with certain implants may be ideal.

Take the example of the synthetic artificial blood valve, the carbon valve. This is one which when it's implanted it is very well tolerated except you have to take an anticoagulant, cumadin, for the rest of you life.

That's the extreme where you don't do anything to make it a material that is potentially non-thrombogenic but you have to provide an antithrombogenic drug. Backing off on that, make the material better, use less drug and so on.

Brash: In other words iterate towards perfection.

Hoffman: Yes, but admit that imperfection is the rule.

Brash: Yes, the admission that the drug regimes are necessary is by itself an admission that perfection is impossible. But we can iterate towards, as you say, less and less drug and better and better surface.

Cazenave: Again, there is diet too. I mean you can eat the right fatty acid from fish.

I mean what you have to achieve is like cooking, this is a French remark. You can buy all the ingredients at the market. Some are able to do well, others not. But everything is there. Maybe we don't use the materials properly, but this doesn't prevent us from trying to achieve better biomaterials.

Baquey: The problem of a permanently implanted material or device is also a question of design. And to take again the example of the cardiac valves, the amount of drugs which are given to the patient are less and less important because of the improvement of the design. The disc is always coated by pyrolytic carbons, but the design is not the same and the patient needs less and less drugs.

And for the vessels, I think that if mechanical properties of the substitutes could be improved in order to get mechanical performances similar to those of the natural vessels, I think that the properties of the surface could be less important.

Lemm: Which performance you mean? The compliance?

Baquey: Probably yes.

Missirlis: But is there any study to indicate the correlation between the improvement of the design of artificial heart valves and the amount of the anticoagulant?

Baquey: Yes, I think that there are epidemiologic studies, which show that the St. Jude valve, for instance, need less anticoagulant and less antiplatelet drugs than the Bjork valve or other disc-based valves.

Missirlis: Could I rephrase that? Was the better design designed for this purpose or this is an offshoot of a better design? You understand my question?

Hoffman: I can give you an answer. The rational for the Saint Jude valve is to attempt to take this very unnatural material, this pyrolytic carbon deposited on a graphite support, and create a valve that allows you to have central flow, such as you have in a natural valve.

The Bjork-Shaley disk has flow in two ways and it's not nearly as natural as the Saint Jude. That's the purpose. If a man imitates nature in that way, approaches something more compatible, at least in this case.

Missirlis: I was in a recent meeting on heart valves standardization methods of EEC and most people in Europe use Bjork Shaley rather than Saint Jude. How is it in the States?

Hoffman: I don't know the statistics.

Missirlis: In Europe it seems that almost one or two centres use only Saint-Jude.

Hoffman: Let me tell you an interesting comment. It needn't go on record but nevertheless it's interesting.

I occasionally watch the stocks on television and the Saint-Jude company stock has been going up very rapidly. They are selling a lot of valves.

Brash: I have talked to people in the heart valve business and one trend seems to be towards bioprosthetic valves rather than 'mechanical', valves.

Another thing they always tell me is the use of any particular valve is very much a function of what the surgeon started with and if he feels comfortable working with one valve and has reasonably good results he's going to stay with it.

A new surgeon coming into the field might be able to decide logically, having looked at the evidence, what he's going to use. But I'm not sure that kind of information is really available at this point.

Missirlis: Well let's not divert the discussion, because I could ask many questions on that. But as we were talking I though if one had all the people available, I mean blood chemists, physical chemists, surgeons etc., could one device a scheme of experiments and methods to lead to the ultimate answer? I think one could say something or make a comment on that.

Beugeling: May I make a comment. I don't think so at this moment. Everyone is trying different things. I want to compare it with a situation in the 1920's. If you had given a lot of money to the ladder builders and rocket builders they would not have reached the moon, because the time was not yet ripe for it. I think we have a similar situation at the moment.

Hoffman: In a sense I think the devising of a scheme is basically listing a series of hypotheses that you believe are valid to test.

We've already identified a number of them. One is that a surface that is totally repellent to proteins may well be repellent to platelets, and therefore may be an ideal blood compatible surface. You can write a grant on that.

Another hypothesis is that you want a surface to adsorb physically certain key proteins. These proteins you might identify; we talked about albumin and others.

Another hypothesis is that you don't want those surfaces to be allowed to adsorb randomly from plasma. You want to put your own presorbed or precovalently bound proteins on those surfaces. If you're coming again around all these concerns of this meeting.

But then you can go further. You can make a hypothesis that you want a surface that never adheres any platelets. Or that adheres platelets but doesn't activate them for release and aggregation. You can list a whole series of hypotheses. This is very easy. All of them are essentially aimed at the idea that key proteins lead to platelet aggregation and activation of aggregation or thrombus deposition, and that's what you don't want.

Cazenave: I will add that you can identify a hypothesis to test and you need, as you said previously, to work together with people who are more familiar with other areas.

We cannot be competent in everything. We have a real need to talk to each other. I remember many years ago when John and I arrived at McMaster it was very difficult, because we were both speaking a different 'chinese' dialect, and it's difficult to understand the problems of other people.

But by doing that, obviously it's always the same in science, we plan to do research, which doesn't mean that we are going to find something. There is a difference between research and discovery. Planned research doesn't lead very much to major discoveries. You need a little bit of luck.

Brash: I like Allan's summary of valid hypotheses and how we should proceed, but I would add to his list perhaps that we might want to encourage fibrinolysis on surfaces by some means: perhaps by preferential binding of plasminogen, or tPA or some such thing as that.

That's another strategy which really we haven't touched on here and which is a favourite one of mine.

I think that Allan has named a few strategies which could run parallel to each other and not necessarily be interconnected in any way.

I think your original question implied that an assault on the problem could be planned in the manner of what used to be called a PERT chart, with various tasks assigned to different groups. These would interact along the way and finally everything would converge to the final answer.

We are not in that situation at the present time. I think we have a number of threads running parallel to each other but there is no sign at this point that convergence is going to occur. We can only hope.

Baquey: I think that another thing which has to be paid attention to, in order to develop blood compatible materials is related with the use of devices in extracorporeal circulation; in this situation the major drawback of the device, apart from the thrombotic phenomena which is well controlled by heparin therapy, is activation of the complement.

And the question for a blood compatible surface is to know its ability to activate or not some fragments, especially the C3 fragment, of the complement system.

Hoffman: I think it's difficult to develop all the consequences of such an activation, but we know that this fragment is as much present as fibrinogen, near as much 1.6 mmol/ml in serum or in the plasma. A few people pay attention to the adsorption or the behaviour of such a protein in front of the surface.

I think that complement activation and the resultant disappearance or dropping concentration of a number of white cells is a very important clinical problem. Here is an example of a company, when testing an hypothesis ended up with a new product. They are identified as 'Baxter health care'. I think it's well known that the surfaces with polar groups that are not charged (like amino groups or hydroxyl groups) are good activators of a complement. Surfaces that have charges, particularly negative charges, seems to resist the binding of these active factors and continue the activation of the sequence.

So what did they do? They took cellulose, the most common membrane used in dialysis and they made it negatively charged by treating it with acidic anhydrate. It's so simple. And then they produced acidic acid groups, COOH groups on the surface. Now they have a negatively charged surface. A very simple treatment a cheap treatment, one you can do in large scale.

It's a kind of a model-I don't know if it's clinically up – but it's a model that we can think about to test the mechanism on a molecular level hypothesising a chemical change to a polymer surface and test it again. And this one seems to work.

Brash: I'd like to follow up on that based on a discussion that Charles Baquey, Herbert Jennissen and I had at the coffee break after having discussed the question of 'did research on protein adsorption lead to any good surfaces', when we were unable to come up with anything very definite in the realm of non-thrombogenic surfaces.

But clearly in the case of solving the problem of complement activation, research on protein adsorption did in fact lead to 'complement resistant' surfaces. That should be put in the record on the previous discussion, as a very solid and definitive example of that type.

Hoffman: In fact, the stimulus for this was a European-based membrane from the Rhone-Poulenc company. The AN69 membrane is a sulphonated membrane with a very low complement activation level.

Cazenave: We have all been talking about what I would call short term problems.

But there is the long term problems. I recall a study in which we were comparing Cuprophane and polyacrilonitrile AN69 for long-term hemodialysis. People who have been dialysed for 15 or 20 years have two major complications: one is premature atherosclerosis, the other is amyloidosis.

We showed in vivo in these people that those who were under Cuprophane showed dramatic complement activation, which is a very rapid phenomenon occurring within less than 15 min after the beginning of hemodialysis. These people have been treated at least 3 days/week, every two days. There was a major difference: the ones under Cuprophane released PDGF mitogen from their platelets and the ones under AN69 did not. And the levels were 3 times greater, so, there are the long term consequences (Ch. 6, Ref. 17).

The compatible surface has also to be a surface which will not release small pertides or proteins which have major general biologic deleterious effect.

Missirlis: So, what do you think about these long term effects? What is the transition time? I mean if a surface is good there, for how long? One week then it's good for ever or is it something that changes all the time?

Hoffman: Can I respond? We worried about the acute or chronic problem in each implant or device that has a different set of requirements.

Take the small diameter vascular graft. We're trying to make a non-thrombogenic surface.

But this may really not be the problem of failure in small diameter vascular grafts. The problem of failure may be what is called an antimonic hyperplasma, which is a thickening of the vessel wall creating what is called a stenosis or a narrowing for blood flow. The less blood you get through there the more likely you have stasis, recirculation pattern and deposition of thrombus.

So we may solve the acute problem, which is avoiding the initial thrombotic occlusion of this small diameter graft, but we still face the chronic hyperplasia, and this may be related to the mechanical properties and not the surface properties of the graft.

So you're right. You have to keep your eye on all sides of the picture, to solve what you can, but you must worry about everything.

Brash: You could also add calcification and mineralization in general to this list of potential long term problems which have so far been identified. I'm sure there will be others in the future if we ever solve the short-term problems that are being worked on at the moment. It's pessimistic, perhaps, to say so but no doubt as we take the layers of the 'onion' off, there is always one more left underneath that we will have to peel off to get to the 'fruit'.

Cazenave: Well, it's not so pessimistic, I think. If you take the hemodialysis problem before the artificial kidney, these people had damaging kidney failure and renal disease and they were going to die very quickly. Now some of them are surviving and can do well, with some problems, because they have been dialysed, some of them, for more than 20 years.

So we could not forecast these problems. This is a new pathology which we have to deal with and it's rather complex, but it's almost amazing that they can be dialysed for more than 20 years.

Baquey: I would like to ask a question to Jean-Pierre, about the calcification of prosthesis: is there any evidence of the role of the activation of platelets with this phenomena related to the osteocalcine content of the platelets?

Cazenave: I don't know.

Baquey: Nothing about the release of this protein outside the platelets during the activation?

Cazenave: Sorry, no.

Koutsoukos: May I answer on that? There have recently been done some studies on that, as far as I know by Prof. Nancolas, in Buffalo, USA on mineralization in the presence of osteocalcine. And they have found out from shearing drop bad systems that osteocalcine first adsorbs strongly on mineral surfaces like HA, hydroxyapatite, and it reduces drastically the rate of calcification. It could effectively zero it.

Missirlis: Well, in one set of questions we had put down, whether protein adsorption studies are comparable to other thrombogenicity tests. What did you have in mind Willy and can we discuss that?

Lemm: Yes, I think we have discussed that. There exist same cases where there is a certain correlation to other tests. So protein adsorption studies have their relevance to predict in a certain way the thrombogenicity of a material, but a clear answer does not exist.

Hoffman: It's still hypotheses. Proteins are involved in thrombus formation, proteins stick to foreign surfaces, therefore there must be a connection. But I don't think you can generalize.

The problem comes down to one of design, or maybe the right drug regime. There are so many other multivariable system factors that it's very difficult to generalize.

However, I'm convinced, that protein adsorption is a fundamental and important step in the interaction of a foreign material with blood. Whether it leads to a thrombus or not may be irrelevant. If you're in a very high flow regime coming out of the aortic valve it may not matter what the composition of that surface is. You may need an anticoagulant anyway but the fact is it may not matter. You're not going to deposit thrombus.

I would guess if you make a smooth surface of any material and use that as the disc in the valve coming out of the heart that you would still need an anticoagulant. But you would not see thrombus on that surface. Because the flow is fast, you're going to damage platelets, regardless of the composition of that surface.

Missirlis: But, you see, one reason that some people suggested going into biological valves is to avoid anticoagulation treatment.

But of course then you come to other problems. Like calcification and break down. So, in trying to solve one problem you come up with another. And if you have a child that needs a very at the very early age, a heart valve you can't at least anticoagulate him for ever.

There is some evidence, I think that there is some other organ damage with long-term anticoagulation. Is it right or one could use some drugs for ever?

Cazenave: Well, therapy is always a choice. If you can achieve better without drugs, well, most of the time, it's difficult. You have to control your antivitamin K therapy very well.

The lifelong risk of serious bleeding, which is intracranial bleeding, exists, but if the person is very well maintained with a good surveillance as the Dutch have (I think are the best in this respect for anticoagulant therapy) you can do it.

One of the problems is that some of the factors involved in calcification are vitamin K-dependent, so you may have some growth problems. And you have some problems during pregnancy if it's a young girl.

But again I mean it's a matter of choice. No, valve, no life. Valve and anticoagulant you may live for a long time. It's like the hemophiliacs, who died at 15 to 17 years old and were crippled 50 years ago. Now they live a normal life, with difficulties sometimes and apart from AIDS their life expectancy is almost close to normal boys of the same age.

20. Concluding remarks

Missirlis: I think we could spend the next thirty minutes to not really make conclusions, of course we can't make conclusions, but at least say our thoughts regarding all that has been said so far.

And try to remember that one part of the book when we publish it would be to help also young researchers coming into the field. So some recommendations, that, some of them have been already recorded, and I'm grateful for that, would be welcomed. Who wants to say something on this? Let's start with Allan.

Hoffman: I think we all felt that protein adsorption and protein interactions with foreign surfaces are among the most important initial steps in what may lead to a successful or unsuccessful outcome, which is thrombosis and excessive thrombosis. That's something I want to emphasize.

Practically every foreign surface I know induces thrombosis and therefore our objective should be to try to minimize and control this with material modifications and with drugs.

In terms of what we learned here, as advice to the young researchers in the field I would say that it's very worthwhile to understand the interaction of proteins and blood cells with surfaces; that it's very worthwhile to understand the surface you're dealing with; and that it's important to design the right experiment to test this interaction.

That has to do not only with what proteins you select but also with the environment that you study it in, whether you use plasma or whole blood, what the design of the system is in terms of blood flow, and eventually how you try to correlate with other in vitro tests and in vivo or ex vivo tests.

If a young researcher comes away discouraged, I would say to him that there are too many correlations out there that have to be developed. Don't be discouraged because the field is really in its infancy.

We were talking earlier about the artificial blood vessel which is now very successful. This business started with nylon parachutes the Second World War. That's not even 40 years ago.

In the scheme of things in science today this is just an infant idea. And so the

Y.F. Missirlis and W. Lemm (eds), Modern Aspects of Protein Adsorption on Biomaterials,
249-255.
© 1991 *Kluwer Academic Publishers. Printed in the Netherlands.*

whole field of implants and devices is still a great field for fruitful research. I'll stop at this point.

Baquey: Interactions of materials with living tissues and more specially blood are governed by several groups of factors.

And one group is more specially related to what we're calling chemical properties of the surface and this can be well approached through in vitro protein adsorption studies.

But the problem is to relate the results of these in vitro adsorption studies to what can be produced about the behaviour in in vivo situation in which the blood stream has to be taken into account.

So, I would say that the models which are used in vitro to study protein or biological medium interaction with the material have to take into account, as far it is possible most of the mechanical conditions which are valid in vivo. These mechanical conditions will vary according to the type of application which is the objective.

For the in vivo experiments in order to try to go towards better thromboresistant or better blood compatible materials the question is to choose the right model. And Jean-Pierre Cazenave described the limitations of the in vivo experimentation, as far as these ones are aimed to give results which can be extrapolated to the clinical situations.

We have during this session seen a lot of reports dealing with experiments using labelled materials. And I think that as far as beginners are concerned, working with such materials must be done with much care. You must work, as it has been shown, with purified and well characterized proteins and you must check that all the properties, and more specially those which are of interest in the test, are preserved during the labelling procedures and during the whole experiment.

In this respect apart from things that have been already reported, I think that the choice of the supply of radioactive material is also of great importance. I think, for instance, that all the radioactive sodium iodine preparations are not equivalent.

Cazenave: It's easier not to speak the last because everybody is going to exhaust the possibilities.

I think it's difficult to give recommendations. Summarizing, I think that we've been convinced, all of us, that, if we want to progress in the study of biomaterial interactions with biological systems, we have to devote some time to understanding the basic mechanisms of interaction between biomaterials, biochemical species and cells.

By biochemical species we've been talking of proteins. I think they are important, but we should not forget that there exist other things which are called lipids, lipoproteins, and maybe nucleic acids, ions, water. I mean I've always been told that water structure is very important and as I don't know what it is, I feel it's certainly very important and one should pay some attention to it.

We all realise that if we are all here together, that means that collaboration, real collaboration has to be developed between people of different fields in order to obtain insight into what the others are doing, which, for them, is obvious, but not for us and vice versa. This should continue and I think this type of interaction in Europe, as I know it exists in North America, is a good forum for us to go on.

We have come to a conclusion that the thromboresistance or biological tolerance of biomaterials depends on multifactorial causes. I think this is very important and that we also have to put an emphasis on short-term tolerance and long-term tolerance, because the problems are very different and this is important. Well, long-term tolerance needs short-term tolerance, because you have to support your prosthesis for some time before it has time to age, that's obvious.

I'm also very convinced, maybe it's my medical background, that we need hybrid solutions, hybrid solutions meaning that we need people to develop new biomaterials, we need people to look at design and improved hemodynamic conditions. I think this is the key. If you have the best material with bad hemodynamic conditions you ruin the whole thing.

We need good biochemists to isolate purified proteins, we need biologists, and later on we'll need good surgeons once we understand the basic principles. Surgeons who don't only cut and open, it's usually what they think they have to do.

The role of drugs and diet is very important. I think drugs have one short coming, that is that people are able to take a drug for rather a short period of time. Contraception is a very instructive example, it's very difficult to take a pill every day for your life long. You do that for some time then you stop doing that.

Finally, I would advocate that everything is welcome. We need in vitro testing, animal experiments and ultimately and as quickly as possible we need clinical trials to evaluate the materials.

Jennissen: From the protein adsorption side our aim should be to elucidate the mechanism of protein adsorption. Not for only one protein but for many proteins, maybe even all proteins in plasma. Adsorption should not just be studied on a single but on many surfaces. A prerequisite for such a program is to have large quantities of purified proteins from serum including the trace proteins, so that you have sufficient quantities to do adsorption and desorption experiments with.

What we want to learn is a most general mechanism of protein adsorption which will make the number of contact sites (valence), the conformational changes and the time-dependance of these processes predictable.

So technically to make a beginning this means that we have to measure the adsorption and desorption isotherm for each protein and determine the corresponding binding constants.

Finally techniques have to be developed for monitoring the conformational changes about which we know practically nothing. Maybe then can we also begin to understand the binding of cells to the adsorbed proteins.

Beugeling: I want to begin with the remarks of Prof. Jennissen. I also think it is important that you know what a protein is doing on the surface. But if you only study protein adsorption from single protein solutions, you have a totally different situation compared to the one in which protein adsorption takes place from plasma. We already discussed it this morning. You get displacement of proteins. This displacement is time-dependant, that is to say it depends on the time the protein is present on the surface and I think that is very important. Only few people have looked at conformational changes during and after protein adsorption as you said. There are some studies in which single protein solutions were used, but only a few in which proteins adsorbed from plasma. And I think that is one of the basic things, because protein adsorption from plasma is a very dynamic process, it's not a static process in which you can speak of simple adsorption and desorption constants.

Everything has already been said, but I will emphasise that the search for materials which allow the adhesion and overgrowth of endothelial cells is very important. Perhaps other cells must be included.

Brash: I think we all agree it's important to know for a given surface what proteins are adsorbed to that surface from blood or plasma, how much of those proteins are adsorbed (and I don't mean just one protein, I mean as many proteins as are found on that surface), what time sequence those proteins are adsorbed onto that surface in, and what is the physical biological status of those adsorbed proteins.

And I would say, in relation to Herbert Jennissen's comment that a lot remains to be done in that respect. It's a favorite theme of mine to say that we know almost nothing about the composition of protein layers adsorbed from blood onto any material. And we're working on that and I hope that beginners coming into the field will be encouraged to do so also.

I think we also concluded that the Lyman hypothesis is by and large still valid, but we should generalise it to say that it's not just a question of albumin and fibrinogen but of every protein, in relation to thrombogenesis or other biological effects. This leads me to another of my hobby horses, i.e. that a good approach to designing biocompatible surfaces, is to control protein adsorption, and so I would emphasize that.

I think young investigators should be at least exposed to this point of view, whether they accept it or not, and of course examples of controlling protein adsorption, we have already gone through: zero protein adsorption or the right proteins or the right combination of proteins.

Future things to work on in addition to those I've already mentioned, which I think have been focused by this workshop? One, possibly the most important one, is that we need to do a lot more on correlation between protein adsorption experiments per se and thrombogenic events which may be correlated to protein adsorption. And I mean that for in vitro experiments as much as for in vivo.

As the group here has considered this question we've come up not quite empty handed, but almost. And we recognise that there is work going on at the present moment which is not yet finished and which we can't make any conclusions on, but I think that people should be encouraged to move in this direction.

Baszkin: I agree with all of you. As a surface chemist I would like to add one more remark. I'm convinced that we need more direct measuring techniques to be able to probe different events that occur at the interface between biological liquids and polymer surfaces.

I wish to point out that I'm particularly convinced as to the validity of Dr. Jennissen and Dr. Brash remarks in regard to some specific aspects of protein adsorption phenomena.

I believe that the development of different in situ measuring techniques, designed for characterization of polymer surfaces, polymer-protein and polymer-cell interactions is of an extreme importance to make a decisive progress in the understanding of the in vivo behaviour of biomaterials.

Van Damme: Maybe one small comment: I heard these days a lot of advises for beginning researchers and I think if I started today I would have done a lot of things different. So I hope that the information from these days will reach the beginning researchers, maybe through the book.

Magnani: These days has been very useful for me, as well as for the other beginning researchers, I think. I learned a lot of things and I took note of the several problems, too. Now, I hope all that has been pointed out today, tomorrow it could be done.

Koutsoukos: I would also like to add something to all things said, that I think that the surfaces we use as substrates have to be looked upon as charged interfaces as they are.

And therefore emphasis should be placed when we look at adsorption not only on the protein part but on the substrate part as well, especially on the aspect of the electrical charge that the interface has in biological fluids.

This, as we said before also, is possible to affect not only the extend of proteins adsorbed but also the conformational the proteins have on the surface, and it may also affect the adsorption of co-ions, because once we have proteins adsorbed we may have co-ions that maybe important.

Finally I would like also to say that protein adsorption is also important in the long range, in terms of calcification as it was mentioned before, and calcification maybe retarded or maybe in some cases promoted depending on which ions are attracted on the surfaces of the biomaterials.

Because it's believed now, it's widely spread the first nuclei maybe formed by complexes of calcium and phosphate that exist on surfaces which may adsorb also from blood.

254

Lemm: I am no longer working on this field, so my statement may not be up to date regarding the scientific level. But I have to confess that I learned a lot in asking all these questions.

And this meeting reminds me in a way of a story dating back to the time of the ancient Greeks. Not far from here lived a dangerous monster, the nine-headed Hydra. No one was able to overcome this threatening creature. Whenever one head was cut off, two new ones appeared immediately. So, the beast became more and more horrifying, up to the day when Herakles came and solved this problem...

This phenomenon seems to be transferable to this workshop:whenever a question was asked two or more new questions arose. Herakles burned the wounds and thus no more heads could appear.

Obviously the 'monster' protein adsorption is not that easy to conquer. Like a Herakles we simply have to find better tools and the key to cunning ideas.

Nevertheless, I would like to thank you for participating and I would like to express my deep gratitude to everyone, especially to those who came from the United States and from Canada. We learned a lot from them and I hope that they did profit also from us. Thank you once more for coming and contributing.

Hoffman: In fact as we went along and I began to right done ideas and pretty soon I noticed that most of these things have been covered by other people.

One thing I just bring out: I don't ignore the idea of immobilizing biologically active molecules on surfaces. I think that's very important. Be critical and observant. Observation is the mother of invention. And be critical. Because this field is loaded with people who reach conclusions too fast. So both of these you must remember.

Do a battery of tests. This is another idea I think was mentioned many times. One test is not going to do it. Many tests will do it.

And I just want to congratulate you two guys. Because this was a wonderful idea. I thought – John Brash and I we've talked and we've never been in a meeting like this before. Usually you come to a meeting and you sit and listen to other people give papers and you get up into yours and then you sit down and sit to listen some more. This time we worked the whole time. There was never a change to dose or to think of anything else. It's a great idea. I think you handled it beautifully and Yanni – we shouldn't let Yanni get away without telling us his conclusion.

Missirlis: I excepted that. Well, I'm not in biomaterials. It's something that I've been courting it for some time for the simple reason that I've been working with heart valves for the past 15 years and that's a biomaterial, in one sense, natural biomaterial.

I've been working with erythrocyte cells and that's also a natural biomaterial and for one reason or another, it seems that every time these cells and tissues come into contact with some artificial biomaterials.

So, for the past year and a half, I've been thinking of going into this area, because in Greece, as far as I know, nobody is working in this field. And my idea is to get as much information as possible. I've George Michanetzis who is going to do work in this area, going to various labs in Europe, trying to learn techniques, trying to learn methodology, what is important and all this information we'll try to digest and hopefully to establish some nucleus of lab in our Biomedical Engineering Laboratory.

I think all this has been very helpful for me and George and my other colleagues and I'm really grateful for all the comments that have been heard.

I promise to do as much as I can in decifering all these out of the tapes and I'll be in contact with most of you or rather all of you most of the time in the next few months in trying to get a book out in print as soon as possible.

Many questions from the fluid mechanics point of view which is considered important, I think we didn't address ourselves to that very much, mainly because other things took up most of the time and the interests of the people probably were directed in that sense. I have an interest in fluid mechanics part and I think in my mind some tests might be – forming in situ testing methods where fluid mechanical environment should have it's importance.

Well, having said that, it's my turn to thank each and everyone of you who contributed to this workshop. It has really been a great pleasure for me to see people that I've met 20 years ago and some that I met only a few months ago here. And I hope to see each other pretty soon some place. So, we may conclude at this time.

Authors' index

Baquey: 96, 100, 104, 105, 107, 113, 121, 123,
125, 144, 145, 146, 152, 154, 158, 159, 160,
163, 164, 167, 189, 190, 224, 225, 226, 227,
231, 234, 236, 240, 242, 243, 245, 247, 250

Baszkin: 101, 103, 111, 113, 114, 115, 116,
117, 118, 121, 122, 124, 126, 127, 128, 132,
133, 139, 141, 142, 145, 151, 154, 157, 158,
159, 160, 175, 186, 193, 207, 217, 223, 224,
225, 230, 234, 237, 238, 253

Beugeling: 95, 96, 100, 102, 125, 130, 131, 140,
143, 145, 146, 147, 151, 154, 156, 158, 163,
169, 173, 174, 175, 176, 196, 199, 201, 202,
207, 209, 210, 219, 222, 223, 224, 225, 232,
234, 236, 239, 241, 243, 252

Brash: 95, 96, 97, 98, 99, 100, 101, 102, 103,
104, 105, 106, 107, 108, 109, 110, 111, 114,
116, 117, 118, 119, 121, 123, 124, 125, 127,
128, 129, 130, 131, 132, 133, 134, 137, 141,
143, 145, 146, 147, 151, 152, 153, 154, 155,
156, 157, 166, 167, 168, 173, 175, 177, 179,
182, 183, 184, 186, 187, 188, 189, 190, 191,
192, 193, 195, 196, 197, 198, 201, 203, 204,
205, 206, 207, 208, 212, 213, 214, 216, 219,
221, 222, 223, 227, 228, 232, 233, 234, 235,
236, 237, 239, 240, 241, 242, 243, 244, 245,
246, 252

Cazenave: 96, 97, 98, 99, 100, 101, 102, 103,
105, 106, 107, 108, 109, 110, 115, 118, 120,
126, 127, 130, 133, 134, 140, 141, 142, 143,
165, 167, 186, 188, 189, 194, 196, 199, 204,
205, 206, 213, 214, 219, 222, 224, 225, 231,
232, 235, 236, 238, 239, 240, 242, 244, 246,
247, 248, 250

Hoffman: 95, 98, 99, 101, 104, 106, 108, 110,

115, 116, 117, 119, 126, 127, 129, 130, 131,
135, 138, 139, 141, 142, 143, 146, 149, 151,
152, 154, 155, 157, 161, 164, 165, 166, 173,
175, 176, 177, 181, 184, 187, 191, 192, 194,
195, 197, 199, 203, 204, 205, 206, 207, 208,
213, 214, 215, 217, 220, 221, 222, 223, 224,
227, 228, 229, 231, 232, 233, 234, 235, 239,
241, 242, 243, 244, 245, 246, 247, 249, 254

Jennissen: 117, 119, 120, 124, 126, 127, 128,
129, 131, 133, 139, 142, 143, 149, 154, 155,
156, 159, 160, 175, 179, 182, 187, 189, 190,
192, 193, 194, 195, 198, 199, 202, 204, 205,
207, 208, 213, 214, 225, 226, 231, 251

Koutsoukos: 155, 188, 192, 247, 253

Lemm: 106, 113, 116, 132, 137, 146, 149, 152,
157, 158, 159, 163, 164, 173, 194, 195, 197,
198, 208, 209, 212, 221, 228, 234, 238, 242,
247, 254

Magnani: 124, 125, 196, 197, 253

Missirlis: 97, 99, 100, 102, 103, 105, 106, 107,
108, 109, 110, 111, 113, 118, 121, 124, 125,
126, 127, 128, 130, 131, 142, 157, 160, 165,
169, 176, 189, 190, 192, 193, 194, 197, 198,
199, 201, 206, 209, 210, 213, 215, 216, 217,
219, 220, 222, 227, 228, 235, 236, 241, 242,
243, 246, 247, 248, 249, 254

van Damme: 176, 177, 203, 204, 207, 208, 253

Wolf: 98, 104, 114, 115, 116, 123, 127, 133,
138, 144, 156, 157, 160, 164, 168, 174, 182,
183, 187, 188, 192, 193, 214, 215

257

Subject index

alpha-1-antitrypsin: 43
acceptor-receptor interaction: 159
accumulation: 145, 179, 224, 225, 227
acid-base interaction: 159
acrylic acid: 158
actomyosin: 198
acylation: 113
adhesive proteins: 50
ADP: 17, 31, 50
adsorption
    albumin: 14, 15, 35, 43, 52, 58, 60, 68, 74,
      75, 85, 138, 142, 164, 172, 196
    fibrinogen: 13, 15, 16, 45, 52, 58, 60, 64, 68,
      74, 76, 142, 144, 164, 172, 184, 185,
      187, 196, 223
    protein: 3, 13, 19, 21, 25, 26, 27, 39, 40, 49,
      55, 63, 81, 95, 106, 121, 149, 164, 170,
      178, 197, 215, 219, 228, 233, 249
    isotherms-kinetics: 23, 35, 40, 61, 64, 65,
      66, 68, 69, 70, 89, 90, 169, 177, 184,
      187, 190, 193, 195, 204, 216
    hysteresis: 68, 69
affinity: 3, 9, 12, 13, 16, 22, 44, 64, 66, 69, 70,
    73, 97, 106, 107, 133, 184, 196, 202, 204,
    219
agarose: 63, 70, 149, 190
alpha-globulins: 95
alpha-granules: 50
AIDS: 100, 248
albumin: 21, 22, 31, 35, 73, 95, 100, 105, 137,
    140, 141, 146, 175, 198, 202, 209, 223, 230,
    239, 241
    labeling: 7
    14-C labeled: 61, 123
    radioiodinated: 12, 26, 123
aliphatic amines: 64
alkyl-residue lattices: 63, 203
alpha-2-macroglobulin: 43
Amide I band: 81, 84, 126

Amide II band: 84, 126, 196
Amide III band: 94, 126
AMP: 31
AN69 membrane: 246
anticoagulant: 4, 233, 242, 243, 248
antiplatelet: 166, 238, 243
antithrombin III (AT III): 4, 5, 9, 13, 14, 43,
    105, 107, 139, 145, 223, 228, 239, 241
    labeled antibody: 11
antivitamin K: 248
apoprotein-A1: 210
apyrase: 31
arachidonic acid: 49
AT III: see antithrombin III
ATP: 50
ATR: 82, 196, 216
autoradiographic methods: 10, 17, 18, 152
A-V by-pass: 12
Avcothane: 77

baboon: 165
BDDS-4: 45
beta-globulins: 95
beta-thromboglobulin: 50
Biomer: 45, 71, 234
beta-lipoprotein: 43
Blomback method: 100
blot: 42, 43, 126
bovine plasma: 78
Brewster angle: 170
BSA: 170, 172
buffers: 103, 104, 202
    Michaelis: 10
    PBS: 32, 88, 119, 138
    Tris: 41, 183
    Tyrode: 31

^{14}C: 19, 21, 113, 115, 116, 120, 128, 129, 131,
    158

^{45}Ca: 152, 158
calcification: 248
capillary perfusion system: 30, 50, 51, 163, 165, 196, 227, 237
captive bubble method: 27, 151
carbamate linkages: 64
carbon-carbon composites: 4
carbonyl diimidazole method: 64
catheter: 50, 166, 242
C3 complement: 42, 43, 50
cellulose-Cuprophane: 25, 28, 41, 42, 52, 150, 246
chain mobility: 56
charge: 56, 59
chemical functionalities: 156, 157
chloramine-T method: 8, 9, 119, 132
chromatography: 96, 97, 107, 126, 192, 240
51-Chromium: 169
CNBR method: 64
coagulation: 39, 49, 55, 103, 141, 142, 224, 225, 230, 239
NA coagulation factor V: 50
cold labeling: 117, 129, 130, 202
collagen: 17, 49, 50, 133, 223
^{14}C labeled: 21, 22
colloidal gold imaging: 153
column bead method: 192
competitive adsorption: 21, 58, 60
complement activation: 49, 55, 143, 144, 245
Compton contribution: 134
contact angle: 27, 28, 57, 149, 150, 152, 158, 215
Coulombic forces: 161
CSTR (continuous Stirred Tank Reactor): 183, 189
cumadin: 242

Dacron: 25, 52, 164, 187, 229, 236
dead volume: 182, 184
DEA Sephadex column: 102
denaturation: 40
dense granules: 50
desorption: 19, 21, 26, 56, 70, 89, 121, 126, 127, 129, 133, 158, 187, 196, 213, 252
dextran: 143
DFP: 130
dialysers: 41, 167, 245, 247
diffusion effects: 44, 192, 201
diffusion limited system: 182, 190, 193, 195, 198
diffusivity: 21, 114, 189, 202
dispersion forces: 160
domain structures: 57
donor-acceptor interaction: 56
DTPA: 11

ECC: 15, 16, 51, 122, 226, 245
ED: 45
EIA: 26, 32, 56, 125, 131, 175, 177, 210
ELISA: 176, 207
electrophoresis: 17, 87, 126, 127, 155
electrophoretic mobility: 87, 88, 89
electron microscopy: 152
ellipsometry: 130, 174, 176, 216
endothelial cells: 25, 28, 52, 140, 235, 237, 239
ESCA: 27, 117, 150, 151, 152, 154, 155, 157, 194
ex vivo: 3, 9, 14, 15, 29, 31, 163, 165, 167, 208, 225, 229, 249

factor VIII: 97, 109, 110, 154, 156
factor XI: 43
factor XII: 29, 30, 36, 43, 222
factor XIII: 52
FEP: 25
fibrinogen: 22, 29, 30, 31, 32, 33, 34, 35, 36, 40, 43, 46, 49, 50, 73, 88, 89, 90, 95, 98, 99, 100, 101, 104, 105, 115, 137, 139, 146, 152, 153, 181, 183, 198, 201, 202, 204, 205, 207, 208, 209, 211, 220, 221, 225, 226, 227, 230, 231, 232, 245
labeling: 8
sources: 96
^{3}H labeled: 61
fibrin: 141, 232, 238
fibrinolysin: 241
fibrinolysis: 49
fibrinolytic system: 102
fibronectin: 25, 29, 31, 43, 49, 50, 88, 95, 102, 104, 105, 223, 236, 237
flow generators: 6
fluidity of cell membrane: 144
fluorocarbons: 229, 239
FPA: 50
free energy: 57, 69
Freundlich type isotherms: 68
FT-IR: 81, 121, 124, 125, 130, 150, 152, 196

gamma counter: 73, 75, 118, 165, 167, 231
gamma-globulin: 73, 223
gas discharge: 150, 164, 229
gelatin: 143
gelatin-coated vascular grafts: 5
gel permeation chromatography: 9
Germanium-GeLi: 9, 85, 196
Gibbs free energy: 69
glass: 34, 36, 89, 111, 181, 185, 196, 201, 215, 220
glutaraldehyde: 97, 142
glycoproteins: 31, 49, 120, 153
Gortex: 229, 236

Gott ring test: 221

Hamer's method: 100
haptologin: 43
HEMA: 25
hemoglobin: 68
hemophilia A: 97
heparinized blood: 51
heparinized materials: 144, 145, 157, 233
heparin-like materials: 4, 11, 13, 14, 16, 145,
    225, 228
heparin-Sepharose: 107
hepatitis: 100
high density lipoproteins (HDL): 26, 32, 33, 34,
    35, 36, 59, 140, 142, 175, 201, 203, 209,
    210, 211, 212, 241
hirudin: 241
HIV infection: 100
^{3}H-labeled protein: 129
HMWK: 29, 30, 32, 33, 34, 35, 36, 43, 59, 105,
    209, 210, 222
hollow fibers: 52, 165
HSA: 35, 51, 58, 61, 88, 89, 90, 187
hydrogels: 221
hydrophilic: 34, 45, 56, 61, 63, 142, 147, 149,
    154, 170, 219
hydrophobic: 40, 56, 61, 111, 147, 149, 154,
    156, 210, 214
hydroxyapatite: 247

^{111}Indium: 11, 12, 50, 128
IgG: 26, 31, 32, 34, 35, 43, 46, 52, 59, 88, 89,
    90, 125, 172, 176, 184, 185, 202, 207, 223
immobilization: 254
in situ: 21, 46, 154, 157, 167, 176, 179, 184,
    190, 253
in vitro: 3, 4, 8, 10, 29, 30, 75, 147, 163, 164,
    165, 227, 232, 249
in vivo: 8, 163, 164, 165, 226, 227, 246, 249
^{125}Iodine: 50, 73, 115, 128, 131, 132, 205, 207,
    231
^{131}Iodine: 132
Iodo-Gen method: 8, 119, 132, 134
ionic interaction: 56
ionic strength: 83
isourea linkages: 64
kallikrein: 29, 30, 36

Kekwick's method: 107
kininogen: 29, 30
Kraton: 77
Kusserow ring test: 221

labeling methods: 7, 8, 9, 12, 16, 17, 19, 113,
lactoperoxydase method: 8, 119, 132, 134

laminin: 236
Langmuir-type isotherm: 65, 68, 160
LDL: 211, 241
Lee-White test: 220, 222
Lindholm test: 221
loosely-bound: 133, 207
Lyman hypothesis: 73, 137, 139, 142, 146, 223,
    234, 252
Lyman test: 221

macroglobulin: 42
MDA: 45
mercaptans: 64
methylation: 113
MMA: 25, 57
molecular mobility: 150
mongrel dogs: 5

non-fouling surfaces: 228, 239
Nose chamber test: 78, 222

PAMA: 56
PDGF mitogen: 246
PE: 4, 5, 11, 31, 32, 35, 111, 115, 133, 147, 158,
    165, 209, 210, 211, 212, 223, 225, 241
Pellethane: 77, 147, 223, 234
PEO: 58, 60, 61, 119, 143, 146, 147, 223, 239
periodate pyridoxal borohydride method: 120
periodate tritiated borohydride method: 120
PETP: 25
PEU: 4, 12, 39, 45, 74, 83, 142, 146, 147, 150,
    152, 154, 206, 223, 234, 235, PF4: 50
PGI2: 30
pH: 40, 83, 193
phosphorylase: 64, 67, 69, 70, 120, 128, 156,
    157, 190, 213
phospholipids: 49, 224
plasmapheresis: 51, 141
plasmin: 102, 103, 204, 232
plasminogen: 43, 102, 104
platelets
    activation: 49
    adhesion: 30, 49, 55, 95, 96, 138, 140, 147,
        164, 221, 223, 224, 225, 227, 237, 239
    aggregation: 31, 49, 55, 96, 97, 140, 225
    agonists: 49, 50
    deposition: 29, 31, 226, 237
    ^{111}In labeled: 12, 16, 17, 30, 50
    morphology: 49
    receptor interactions: 106
    stimulation: 50
platelet derived growth factor: 50
Plathuran: 78
polyacrylonitrile membranes: 52, 246
polyaniline: 155

poly(ethylene teraphthalate): 205
polygalactose: 63
poly(N-vinyl acetamide): 223
polypyrol: 155
polysulfone membranes: 52
porosimetry: 186
potentiometric titration: 155
preadsorption: 31, 51, 52, 88, 95, 138
precoating: 31, 95, 141, 166
prekallikrein: 29, 43
procoagulant platelet factor III: 49
prostacyclin: 168, 237
prostaglandin: 49, 157
protease inhibitors: 102
protein layer
    composition: 39, 55
    thickness: 174
prothrombin: 224, PS: 5, 32, 33, 87, 89, 111, 170, 172, 214
PTFE (Teflon): 25, 26, 27, 28, 33, 52, 151, 187, 229
PVC: 4, 29, 32, 33
PVDF: 41
pyrolized polymers: 158
pyrolytic carbon: 4, 243

radioactivity measuring and imaging: 9, 10, 17, 19
radioisotopic techniques: 3, 8
red blood cells: 30, 138, 144, 195, 226, 237
^{99}Tc labeled: 12, 16, 17
reflectometry: 130, 169, 173
refractive index: 174, 177
renal embolus test: 221
Re number: 169, 199
reversible adsorption: 19, 21, 40
rigidity: 64
roughness: 150

scintillation camera: 9, 17
SDS-PAGE: 41, 42, 96, 98, 107, 110, 119, 126, 128, 138, 189, 208
SEM: 32, 50, 149
Sepharose: 63, 67, 69, 96, 11, 102, 154, 156, 240
sialic acid: 120
Silastic: 77
silicon: 4, 74, 77, 89, 106, 147, 150, 187, 220, 223
SIMS: 150, 152, 154, 157
smooth muscle cells: 140, 237
sodium azide: 73, 109
spacer arms: 156, 159
specific catalytic activity: 127
specific radioactivity: 127, 132

sucrose: 174
sugar: 174
supply limited system: 182
surface activation: 30
surface mobility: 57
surface potential: 138

^{99}Tc: 7, 12, 118
TCPETP: 25
TCPS: 25, 28, 52
tetrafluoroethylene: 230
thawing: 109
thrombin: 4, 50, 96, 101, 103, 104, 139, 143, 166, 168, 225, 231, 238, 239, 241
thrombomodulin: 239
thromboplastin: 222
thrombospondin: 49, 50
thrombus formation: 39, 49, 78, 164, 219, 227, 234, 241, 248
TMAEMA: 25, 57
topography: 150, 157
total internal reflection technique: 175
transferin: 43, 143
transglutin: 52
transition electron microscopy: 152
transport limited system: 190, 198, 216
trichloroacetic acid precipitation: 208
tunnel electron scanning microscopy: 152
turbulent flow: 199
tyrosine residues: 119

ultrashort time tests: 201, 208, 217
unstirred layer: 190

valves: 243
van der Waals forces: 68, 160
vena cava ring test: 221
vitamin K: 143, 248
vitronectin: 104
VLDL: 211
Vroman effect: 23, 42, 46, 173, 176, 184, 201, 202, 203, 204, 205, 206, 207, 208, 210, 216, 217
vWF: 29, 31, 49, 50, 95, 98, 101, 102, 104, 105, 139, 223

washed platelets: 51
white cells: 144
Wilhelmy plate technique: 56, 114

X-Ray crystallography: 149

zeta-potential: 57, 87, 187, 198